Dialectical
Materialism
and
Modern
Science

About the Author

1908-1994

Kenneth Neill Cameron was born in England but early moved to Canada where he attended McGill University in Montreal. In 1931 he went to Oxford as a Rhodes Scholar. In the summer of 1934 he went on a trip to the U.S.S.R. From 1934–35 he was Executive Secretary of the Canadian League Against War and Fascism in Toronto. In 1939 he received his Ph.D. from the University of Wisconsin. From 1939 to 1952 he taught English at Indiana University. His first book, *The Young Shelley: Genesis of a Radical* (New York, 1950; London, 1951) was awarded the Macmillan-Modern Language Association of America Prize for the best work of scholarship in 1950. In 1952 he came to New York to work at the Carl H. Pforzheimer Library and published four volumes of the library's Shelley collection manuscripts—*Shelley and his Circle,* 1961 and 1970 (Harvard and Oxford). In 1961 he made a second trip to the U.S.S.R., this time with his wife and daughter, and met with Soviet scholars in his field. In 1973—paperback, 1977—he published a Marxist history of the world—*Humanity and Society: A World History;* and in 1974, *Shelley: The Golden Years* (Harvard), a study of Shelley's later works in prose and poetry. In 1976 he published *Marx and Engels Today: A Modern Dialogue on Philosophy and History.* In 1977 he brought out a volume of poetry, *Poems for Lovers and Rebels* (privately printed). In 1963 he became Professor of English at New York University and later Professor Emeritus. In 1967 he was awarded a Guggenheim Fellowship. In 1971 he received an honorary D.Litt. from McGill University. In 1978 a "festschrift" volume of essays honoring his work—*The Evidence of the Imagination: Studies of Interactions Between Life and Art in English Romantic Literature*—was published by his fellow scholars and former students. In 1982 he was presented the Distinguished Scholar Award of the Keats-Shelley Association of America at the convention of the Modern Language Association of America.

Dialectical Materialism

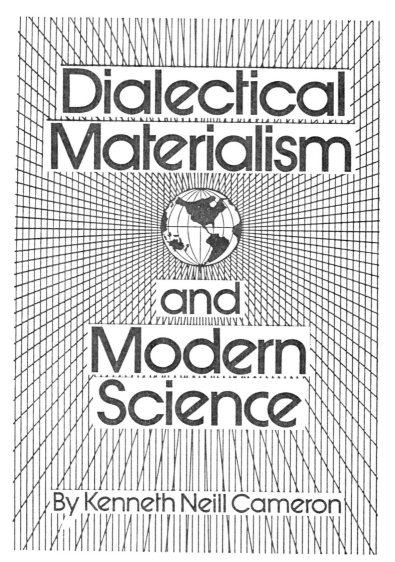

and

Modern Science

By Kenneth Neill Cameron

INTERNATIONAL PUBLISHERS, New York

Library of Congress Cataloging-in-Publication Data

Cameron, Kenneth Neill.   1908-1994
    Dialectical materialism and modern science / Kenneth Neill
Cameron.
        p.   cm.
    Includes bibliographical references and index.
    ISBN 0-7178-0708-8 : $10.95
    1. Science--Philosophy--History. 2. Science--Social aspects-
-History. 3. Dialectical materialism--History. 4. Communism and
science--History.   I. Title.
Q174.8.C36 1994
146'.32--dc20   94-33451
                                                        CIP

CONTENTS

INTRODUCTION

In the last two centuries, thanks to the rise of science, we have been given the tools for understanding the universe and ourselves. We can now begin to see how the universe developed and the particular conditions which enabled conscious life to form on our planet (almost certainly, when we consider the intricate web of evolution, the only conscious life in the universe).

A philosophy based on natural science does not, as the history of philosophy shows, arise spontaneously. Philosophy, like social thought but somewhat more subtly, has been buffetted and shaped over the centuries by conflicting class interests. In time these conflicts produced a thin stream of scientific materialism, first apparently in the commercial enclaves of ancient, feudal India, then in the Athenian and Roman empires, in Bacon's commercial England and Diderot's pre-revolutionary France until with the rise of an industrial working class – new to history – in 19th century Europe, a philosophy based on science could begin to see the general processes of nature and simultaneously render philosophies not based on it obsolete:

> Modern materialism embraces the more recent discoveries of natural science . . . In both aspects [cosmic and biological evolution] modern materialism is essentially dialectic; and no longer requires the assistance of that sort of philosophy which, queen-like, pretended to rule the remaining mob of sciences. As soon as each special science is bound to make clear its position in the great totality of things and of our knowledge of things, a special science dealing with this totality is superfluous or unnecessary. That which still survives of all earlier philosophy is the science of thought and its laws—formal logic and dialectics. Everything else is subsumed in the positive science of nature and history.

So, Frederick Engels, writing in 1877.

In order to understand nature, Engels argues, all we need is natural science and the ability to examine rationally, using both deductive logic and the more complex thought patterns of "dialectics," by which he meant fluid thinking that takes contradictions and interconnections in natural and social process fully into account. Any philosophy that imposed ideas upon reality, as most of them did instead of deriving

1

them from reality, was simply creating delusion. There is nothing except nature and society, no supernatural forces, no God. If the ancient materialist stream, cascaded by Marx and Engels to include social as well as natural science seems to some sluggish at the moment, it will, as I suggest in Chapter Eight, soon be swollen by the intertwining floods of natural disaster, capitalist crisis and a new socialist advance.

Since Engels wrote, science, of course, has made great strides. Our ideas of matter have been reshaped by elementary particle physics and our ideas of the universe by astrophysics. We have uncovered connections between matter and living matter and learned something of the secrets of heredity. New sciences have arisen – molecular biology, brain research, sociobiology, ecology and climatology among others – that have thrown new light on what we are, where we came from and where we may be going. The philosophical impact of today's science on today's "modern materialism" I discuss in the central chapters. I begin with a sketch of philosophy, particularly materialism – Lucretius, Bacon, Diderot – and then examine the dialectical materialist views of Marx, Engels and Lenin. In chapter eight – Life, Death and Purpose – I explore the implications of dialectical materialism for personal life and values.

Chapter One

THE COURSE OF PHILOSOPHY

Although philosophy arises from the confrontation of humanity with nature, internal and external, its course is determined by the course of history. It began with human beginnings in hunting-gathering society and assumed written form some 5,000 years ago in the feudal civilizations of Egypt and Sumer (the epic of *Gilgamesh*). As similar societies arose in India, Persia, China and elsewhere philosophy developed – Gautama to Confucius – as society developed. When the slave-commercial civilizations of the northern Mediterranean emerged they produced new imprints – from Plato to Lucretius. With the "fall" of Rome, philosophy reflected the thinking of the new feudalisms of Europe, with their growing commercial component – Thomas Aquinas to Roger Bacon. As this component became dominant in Europe and gave birth to a new society and a new science the resulting turmoil produced a philosophical sweep of unprecedented scope and variety – from Bacon and Déscartes to Kant and Hegel. When Blake's "dark, Satanic mills" became dominant in mid-nineteenth century Europe there came the cultural mutation of Marx and Engels' "modern materialism" (later dubbed by the Russian Marxist, Georgi Plekhanov, "dialectical materialism").

Hunter-gatherers and Farmers

The present fossil evidence indicates that about four million years ago there evolved in Africa apelike but upright prehumans ("hominids") who lived by food collecting and who paved the way for all that was to follow. About 1.7 million years ago there emerged, again in Africa, a people rather like us, but with somewhat smaller brains – *homo erectus* – who by one million years ago had migrated as far as Indonesia and China. They used fire, made shelters, had a variety of tools, did coordinated hunting of large animals; which suggests some language and, hence, some degree of consciousness. They also apparently performed ceremonial rites involving the ritualistic eating of brains (found among later hunter-gatherers also). By 100,000 years ago fully human people

3

existed, people who buried their dead with ceremonies. For at least 90,000 years then, and probably much longer, people have speculated about their general relation to nature. By at least 30,000 years ago a new socio-cultural step upward had begun, as evidenced in Europe by artifacts indicating wide trade, highly artistic cave paintings, calendars showing the phases of the moon carved on bone, and statuettes of women. The advance was evidenced also in the spread of a generally well-developed hunter-gathering society on all continents including North and South America.[1]

By 30,000 years ago, the remains indicate that people lived in houses (not caves) and must have had a family life similar in its general patterns to ours today. They had a variety of weapons and tools – in stone, wood, bone and leather – including axes, knives, chisels and hammers. They had a vast storehouse of knowledge – of hunting, animal life, family life, medicine, boat-building, house-building (sometimes large community dwellings). And they had a lively creative artistic life which must have included dance, music, storytelling, mimicry, as well as sculpture and (cave) painting. Some of these things can be ascertained not alone from the artifacts of the past, but by integrating them with the situation of hunter-gathering peoples living into civilized times, from the German tribes described by Caesar to those observed in the present century by anthropologists. Such comparisons cannot, of course, be valid in all details but they must be valid for basic patterns because, as history and anthropology alike demonstrate, similar social forms in all times and places produce similar cultures. As grave remains indicate, these early peoples had the same kind of beliefs that we find among the Australian aborigines, the Kung people of the Kalihari Desert in southern Africa or the Ona Indians of Tierra del Fuego, on the southern tip of South America, all peoples with roots going back for many thousands of years. The Onas, for instance, believed that the world was created by a supreme god, Temaukel:

> The most powerful of beings, he is pictured as living wifeless and childless in the sky where the souls of ordinary mortals join him; and he punishes with death those who flout his laws. As for creation, he originated the sky when it was yet without stars and the first amorphous earth. Virtually everything else was left to his deputy, Kenos, ultimately a star, who raised the sky, transformed the world, created human beings and taught them Temaukel's code.[2]

Here, as in embryo, we can see philosophy – with a supernatural basis – as it emerged some 30,000 years ago, with roots going back 100,000 or more years. That there were continuities we cannot doubt, but these people knew little of them, for oral tradition shifts and fades like the peoples who create it. As with all hunter-gatherers everything must have seemed mysterious. A great golden fire moved across the bowl of the sky by day, a changing silver orb by night and small, pinpoint fires, some of them in motion – all duly registered in notched bone calendars. The Earth was only the expanse of land or water they travelled for hunting or trade, up the rivers and over the deserts. Cold, flood and storm must have shaken them with terror, especially in eras of glacial-interglacial change.

Alongside the visible world they saw an invisible world, a world of spirits that controlled all aspects of life. In hunter-gathering societies as far apart as Alaska, Africa, Tierra del Fuego and Australia, we find the same basic belief in this spirit world, a spirit for the human body, for animals, for trees, for the wind, and a supreme sky spirit, all of whom the shamans (ur-priests) attempted to influence by magic ritual. It was worldwide. And it had roots in fear, particularly the fear of death. After a death, the Australian natives "wail and sob, gashing themselves with spears and knives, axes and sharp stones until the blood flows." A tabu forbids the future mention of the dead person's name. And in these and other "fringe" peoples, we find beliefs in a kind of vague ghostly survival, sometimes in the bosom of the sky-god.[3]

Anthropologists tell us that in hunting-gathering societies, all people share these beliefs. This may be so but it may also mean that it is as dangerous to express doubts in shaman-ridden societies as in Church-dominated ones. The skepticism that we find in the first civilized societies certainly had roots, some of them perhaps going back for many thousands of years. But whether they shared the general beliefs of the society or had skeptical reservations, these early people had no realization of their situation. They were, in fact, almost certainly the first and only conscious beings in the universe. The animals they hunted were their evolutionary ancestors and they themselves had come into being as the result of an almost infinitely improbable sequence of events. But of all this they knew nothing.[4]

By some 15,000 years ago, farming societies had begun to grow in some areas, arising, perhaps under pressure of climatic change, from the peripheral animal-and-crop-raising of hunting-gathering societies. Although with the coming of farming communities, societies became more stable – for it was no longer necessary to follow animals in their migrations – and better organized, socially and politically, the basic lack of control continued, and with it the all-penetrating fear that could be used for rule. Hence, fantasy still reigned even though the sky-god of the hunters was supplanted by various agricultural deities and small replicas of oxen took the place of hunting weapons in graves.[5]

Early Fuedal Civilizations From *Gilgamesh* to Confucius

The first civilized societies, so far as we now know, arose some 5,000 years ago in Egypt and Sumer (located in the lower reaches of the Tigris and Euphrates rivers in what is now Iraq). Of these, Egypt was a centrally organized feudal state in which the great mass of the people were peasants working on land owned by great-landowners. Industry – mines, quarries, forests – and transport were worked by slaves. And in addition to the greatlandowners, peasants and slaves, there were merchants and artisans, some of whom ran workshops.

Between the greatlandowners and the merchants there were clearly areas both of common interest and of conflict. The merchants helped the sale of the landowners' products but the landowners, as in all feudal societies, feared their rise and held them back. It was in the interest of both classes to keep the peasants and slaves in bondage but more to the landowners' interest. Hence there were both distinct and overlapping areas of ideas.

The peasants and slaves doubtless, as in later similar civilizations, retained many of the "animist" views of nature inherited from past societies and these were integrated – as also in later similar civili-zations – in part into a State religion set up by a section of the landowning class organized as a Church. Sumer, unlike Egypt, with its centralized states, consisted of a series of independent city-states strung along the rivers which, although surrounded by feudal farming areas, were primarily trading and manufacturing centers. In Sumer, but not in Church-dominated Egypt, there arose a group of intellectuals who wrote poetry, philosophy and social commentary. As the social commentary makes clear, the group in Sumer expressed skeptical social views which must, as in later such societies, reflect those of the merchants, artisans and professionals in their opposition to the feudal agricultural interests (an opposition that appears to have resulted at one point in the commercial interests taking over one of the Sumerian cities); such views, for instance, as that embodied in the cynical verse:

> You go and carry off the enemy's land,
> The enemy comes and carries off your land.

The same skepticism permeates the philosophical views of the Sumerian epic, *Gilgamesh,* the first major work of written literature and philosophy alike.

Gilgamesh was a historical figure, king of the city state of Erech. In the poem, he is shocked into a search for the meaning of life by the death of his close friend Engidu (beautifully rendered in English verse by William Ellery Leonard):

> *What kind of sleep is this*
> *that hath now seized upon thee?*
> *dark is thy look,*
> *and thine ears take not my voice!*

But his search proves futile, and the poem develops a sophisticated skepticism similar to that of Voltaire's *Candide* some 4,700 years later; it is useless to seek for a meaning to life, all we can do is enjoy what it has to offer:

> *Gilgamesh, whither are you wandering?*
> *life, which you look for, you will never find,*
> *for when the gods created man, they let*
> *death be his share, and life*
> *withheld in their own hands.*
> *Gilgamesh, fill your belly —*
> *day and night make merry,*
> *let days be fully of joy,*
> *dance and make music day and night.*
> *And wear fresh clothes,*
> *and wash your head and bathe,*
> *Look at the child that is holding your hand,*
> *and let your wife delight in your embrace,*
> *These things alone are the concern of men.*

There is no purpose to life, and no life after death:

> *What was left over in the pot,*
> *what was thrown in the street,*
> *that is thy fate.*

With no hope of finding an answer to philosophical problems, life's only meaning lies in life itself – in sexual love, in children, in sensuous pleasure, in the beauty of art. As for "the gods," we learn in another poem:

> *When a sky above had not even been mentioned*
> *And the name of firm ground below had not been thought of;*
> *When only primeval Apsu, their begetter,*
> *And Mummu and Ti'amat—she who gave birth to them all—*

Were mingling their waters in one;
When no bog had formed and no island could be found;
Had been named by name, had been determined as to his lot,
Then were gods formed within them.[6]

The gods, then, did not create the world but were themselves created by the same "primeval" forces that had created it. (Apsu was river or spring water, Ti'amat the sea, Mummu clouds.) Similarly in an early Egyptian text we learn that "Ptah, the Great one; he is the heart and tongue of the Ennead of gods ... who begot the gods." However, although the nature of the prime creative force in the Sumerian text is material, that in the Egyptian appears to be spiritual ("heart and tongue"). The implications, then, in some Sumerian and perhaps also in some Egyptian literature are anti-religious. The wealthy prelates described in Sumerian poetry – with oxen to plow their "onion fields" – can hardly have relished being told that the world sprang from natural forces, that there was no immortality and no discernable purpose to life. The social roots for this anti-religious skepticism must have been the same as those of the political and social skepticism, namely the middle class of merchants and artisans, who not only had economic interests conflicting with those of the greatlandowners and their church but had no threatening mass of exploited peasants to dope.[7]

Gilgamesh, of course, was a king, and Engidu, whom he mourns, the companion of a king. In an earlier scene the poet depicted slaves making arms for the two warriors. Gilgamesh would not have mourned the death of a slave. Nor would the poet have considered it a worthy subject for philosophical searching. Gilgamesh is allowed the luxury of skepticism so long as it stays within upper class confines.

The intellectual leaders of the first civilizations doubtless included those of other (smaller) states as well as Egypt and Sumer, but all were apparently centered in the east Mediterranean and Persian Gulf areas. (Chinese civilization, for example, was not to arise for another 1,500 years.) Their intellectual level was high, as is clear from *Gilgamesh* alone. So, too, the esthetic:

Where words are flung out in the air but stay
Motionless without an answer.

The Sumerians and Egyptians had considerable industry – mining, quarrying, lumbering – and extensive trade (with bank branches in foreign lands). They had great palaces and temples. They lived in many-servanted luxury on the banks of the Nile or the shining canals of Sumer They had organized political systems, with parliaments in some

Sumerian cities. They had large and disciplined armies. But some things they could not control: the storms and floods that ruined crops and canal systems or revolt from within or war from outside. The destruction of the city of Ur by armies from Elam (Iran) was vividly described in several poems, its gaping walls and sunken ships and the bodies of its inhabitants piled high in the streets:

> In all the streets and roadways bodies lay.
> In open fields that used to fill with dancers
> They lay in heaps.

The war ravage was compared to a storm:

> The storm ordered by Enlil in hate, the storm which wears away the country,
> Covered Ur like a cloth, enveloped it like a linen sheet.

Although the Sumerian rulers did not literally believe that the god, Enlil, had caused the war – they had themselves deliberately made war – there is clearly an element of terror in the poem arising not only from the destruction itself but from confronting the uncontrollable. Such fears, that is to say, to some degree, affected the very ruling classes who propagated them for mass narcotics. Like Ur in the storm of battle, fear must at times have covered them "like a cloth" of death. Yet skepticism existed – as *Gilgamesh* testifies – and it may not have been restricted to the upper classes, a skepticism perhaps sometimes linked with rebellion: "Not all the households of the poor are equally submissive."[8]

Between early farming society and civilization came a more advanced farming society, with small towns, widely organized trade, specialized manufacture, and class divisions. By 7,000 B.C., 4,000 years before Sumer and Egypt, there was a walled town at Jericho whose population has been calculated at 2,000. Another town, in what is now Turkey, had a population of perhaps 6,000. Graves reveal upper- and lower-class burials. Murals and figurines indicate the existence of other arts – music, dance, oral literature. The remains of temples of about 4,000 B.C. show the rise of organized religion. Much, then, of what has been considered the creations of the Sumerian and Egyptians in the arts, philosophy and social thinking had roots going back in the same general area for several thousand years.

The early Egyptian and Sumerian ruling classes inherited a set of accepted beliefs from later farming societies which they developed in the social whirl of advancing feudalism. In time, as the feudal order took form, they had a generally accepted establishment outlook, social and philosophical, hallowed by tradition.

In addition to class oppression, Engels in his *Origin of the Family, Private Property and the State,* noted another oppressive force, that of the domination "of the female sex by the male." This force, as Engels did not make sufficiently clear, acts in two ways, through social structures – economic exploitation of women, prostitution, anti-feminist legislation – and through individual males in personal, family or group relationships: rape, wife-beating, incest. Thus we have, cutting across those of the class struggle, the currents of a male-female struggle, much of which is, indeed, as a socialist-feminist writer has argued, "hidden from history." Nevertheless it has been constantly present, and it is apparent in philosophy and religious practice: in the argument that women have no souls, in the subordination of women in Church hierarchies, and in more subtle ways, such as the advocacy of celibacy and asceticism (the woman as morally "evil"). Early in *Gilgamesh* we are introduced to a temple prostitute. We find the Church in these early, as in later, feudal orders supporting the social subordination of women.[9]

Following Sumer and Egypt, a major civilized state arose in northern India, along the Indus River. Archaeological remains show that this civilization, too, was basically feudal, with greatlandowners and peasant labor, palaces and hovels; but also with a strong commercial component. The Indus, like the Nile or Tigris-Euphrates, served the dual purpose of irrigation and trade. We have no written records from this civilization but we do from a similar one that arose by conquest on its foundations. In the *Rigveda* poems of about 1500 B.C., we find similar skeptical and materialist outlooks to those in the Sumerian poets:

> Darkness there was: at first concealed in darkness, this All was
> indiscriminated chaos.
> All that existed then was void and formless: by the great power of warmth
> was born that unit . . .
> Who verily knows and who can here declare it, whence it was born and
> whence comes this creation?
> The gods are later than this world's production, who knows, then, whence
> it first came into being?

Once again the "gods" are later creations and the original creative force is material – "warmth" (presumably the warmth of the sun). And skepticism prevails.[10]

Although in Sumer we find both religious views – the gods – and the beginnings of materialism, there is no "idealist" philosophy attempting to establish a reasoned base for a religious outlook. This comes first, apparently, with Hinduism, the first recorded idealist (idea-ist) philosophy, which asserted the primacy and ubiquity of the divine Idea in the universe and life. Hinduism, with roots extending back to the *Rigvedas,* was well established by 600 B.C., Indian feudal society having changed little in the long, intervening centuries. According to Hinduism, the world was created by the god Krishna, who was indivisible, changeless and immortal, in contrast to matter, which was divisible, changing and perishing; the human mind or soul was akin to Krishna, the body to matter; the soul could enter the realm of Krishna only through a series of transmigrations. Hinduism, with its full written records, gives a sophisticated picture of the social function of religious doctrine. In the Indian epic, the *Mahabharata,* for instance, when the prince Arjuna hesitates to go into battle, Krishna explains matters to him in terms of a mystical "self" (soul):

> These bodies are perishable; but the dwellers in these bodies are eternal, indestructible and impenetrable. Therefore fight, O descendant of Bharata!
>
> He who considers this (Self) as a slayer or he who thinks that this (Self) is slain, neither of these knows the Truth. For It does not slay, nor is It slain.
>
> This (Self) is never born, nor does It die, nor after once having been, does It go into nonbeing. This (Self) is unborn, eternal, changeless, ancient. It is never destroyed even when the body is destroyed.

One should not, then, fear death in battle because the soul is immortal. And the ruling-class leader should not hesitate to lead the masses to destruction for immortality conveyed moral sanction for slaughter: "The dweller in the body of everyone is ever indestructible; therefore, O Bharata, thou shouldst not grieve over any creature."[11]

Hinduism also developed the first sophisticated doctrine of immortality. It goes beyond the traditional view, rooted in hunting-gathering societies, of the bodily spirit surviving after death and presents the concept of an eternal spiritual substance weaving its way through all the forms of life and ascending within and beyond them.

Arising apparently at about the same time in India as Hinduism, the philosophy of Mahavira and the Jains declared that there was no God or creation, the universe had always existed, matter was eternal, and consisted of "atoms":

The changes of the physical universe are traced to atomic aggregation and disintegration. The atoms are not constant in that nature but are subject to change as development (parinama). . Homogeneous atoms produce different elements by varying combinations.

Matter is connected with time, space and motion. In addition to these there is also spirit, inhabiting plants, animals, earth, fire and so on, of which the human mind is a form. Body and mind interact through a more subtle form of matter. Like matter, space, time and motion, spirit was not created but has existed eternally. On most matters the Jains were skeptical: "We cannot deny or affirm anything absolutely." They were against all killing. Traditionally, they were artisans and small manufacturers, people with a practical outlook, not particularly concerned with the problems of mass rule – of mass doping. They reflect a comparatively highly developed pre-science whose thinkers had argued, long before Copernicus, that the earth was round and rotated on its axis, and had made great advances in mathematics and pre-chemistry, which were later taken over by, respectively, the Greeks and the Arabs. Hinduism, on the other hand, like similar mystical views in other feudal states, clearly reflected the interests of the greatlandowners whose church, government and army arose on a foundation of peasants and slaves.[12]

Siddhartha Gautama, known as the Buddha (the enlightened one), the most famous of the Indian philosophers (c. 563–483 B.C.), was an anti-religious skeptic. Like the *Gilgamesh* poet, nearly 2,000 years before him but living in a similar social order, of feudalism with commercial enclaves, he argued that the basic philosophical questions were unanswerable:

Accordingly, Malunkyaputta, bear always in mind what it is that I have not elucidated, and what it is that I have elucidated. And what, Malunkyaputta, have I not elucidated? I have not elucidated, Malunkyaputta, that the world is eternal; I have not elucidated that the world is not eternal . . . I have not elucidated that the saint neither exists nor does not exist after death. And why, Malunkyaputta, have I not elucidated this? Because Malunkyaputta, this profits not, nor has to do with the fundamentals of religion, nor tends to aversion, absence of passion, cessation, quiescence, the supernatural faculties, supreme wisdom, and Nirvana; therefore I have not elucidated it.

Gautama, then, was noncommittal on whether or not a creative god existed or whether there was life after death, questioning both the

Hinduist "dogma" that matter was created and the Jainist "dogma" that "the world is eternal," as well as the belief they both shared on the existence of a soul or spirit. Unlike the *Gilgamesh* poet, however, although he placed the emphasis on life itself, he placed it not on sensual and aesthetic pleasures but on an ascetic and non-activist existence, one that would not disturb the feudal power. After his death, his doctrine, clearly intended for his fellow intellectuals, was perverted into religious mysticism for the mass. To Gautama "nirvana" meant extinction (the implied metaphor is the snuffing out of a lamp); to his later Hinduist perverters it signified heaven. Although philosophically a skeptic, his doctrine was elitist, pessimistic and ascetic: as death is the center of life, the enlightened one should shrink into the passive, sinless self waiting for death. Thus while he attacked religious mysticism and superstition he also preached submission. His ascetic doctrine is also by implication anti-feminist; and anti-feminism is directly implied in some of his comments.[13]

At about the same time that Hinduism and Buddhism were spreading in India, Taoism and Confucianism were developing in China, in a feudal, landowning state, centered at first around the Yellow River. Taoism's fundamental concept is that of Tao ("the way"):

> *There is a thing inherent and natural,*
> *Which existed before heaven and earth,*
> *Motionless and fathomless,*
> *It stands alone and never changes;*
> *It pervades everywhere and never becomes exhausted.*
> *It may be regarded as the Mother of the Universe.*
> *I do not know its name.*
> *If I am forced to give it a name,*
> *I call it Tao, and I name it as supreme . . .*
> *Man follows the laws of earth;*
> *Earth follows the laws of heaven;*
> *Heaven follows the laws of Tao;*
> *Tao follows the laws of its intrinsic nature.*

Unlike Hinduism, Taoism never defines "the way" but leaves it hanging in agnostic conjecture. "Tao," however, like the Sumerian "Apsu" or the Rigvedas' "world production," is clearly anterior to the gods or heaven. The general thrust is anti-religious and skeptical. Taoism also emphasized the interdependence of phenomena:

> *Being and non-being interdepend in growth:*
> *Long and short interdepend in contrast;*
> *High and low interdepend in position;*
> *Tones and voice interdepend in harmony;*
> *Front and behind interdepend in company.*

The circular interactions of opposites inherent in this view was connected with a general doctrine that all the diverse phenomena of the universe came into being through the interaction of two basic oppositional forces, *yang* and *yin*, *yang* being the male principle, and standing for activity, motion and light, *yin* being the female principle, and standing for passivity, rest and darkness. But Taoism also had – as well as male chauvinism – a mystical element: "Confucius and you are both dreams, and I who say you are dreams – I am but a dream myself." Confucius, on the other hand, skeptically rejected philosophical speculation:

> Chi Lu asked about serving the spirits of the dead. The Master said, "While you are not able to serve men, how can you serve their spirits?" Chi Lu added, "I venture to ask about death." He was answered, "While you do not know life, how can you know about death?"

Like his contemporary, Gautama, then, Confucius (551-479 B.C.) was not really a religious leader but a philosophical skeptic and an ethical teacher. Like Gautama also he preached mass passivity: "The relation between superiors and inferiors is like that between the wind and the grass. The grass must bend when the wind blows across it."[14]

That Gautama in India and Confucius in China began to express skeptical and ethically oriented views in the same historical period, views contrary in some respects to those of the established church in both countries, cannot have been coincidental. Although both thought within a general feudal ideological framework, both criticized some aspects of it – perhaps sensing that too severe an exploitation might provoke revolt – and rejected the more blatant aspects of feudal superstition. This must reflect the fact that in both countries at about the same time economic growth had reached a point in which a mainly commercially based middle class had grown sufficiently to present a somewhat dissenting view within the general feudal structure. These views also almost certainly did not arise independently of each other but in cultural interactions based on international trade, which by then was quite extensive.

Slave-Commercial Civilizations:
From Heraclitus to Lucretius

In all the early feudal societies of Asia – from Egypt to China – slave labor was used primarily in the non-agricultural side of the economy, in industry (the mines, forests, and quarries), transport (galleys, wagons), manufacturing – either in large royal workshops or small enterprises, and in households (mostly women slaves). Although the total number of slaves was large, it was a small part of the total workforce, which consisted mostly of peasants working on large estates. The comparative smallness of the slave workforce was due to the fact that non-agricultural enterprise was but a very small proportion of the total economy.

In Athens of the fifth century B.C., this situation was reversed. Athens was a small city-state with little agricultural land, a large coastline and good harborage. What farming it did – by peasant labor – was mainly commercial: olives and grapes, made into olive oil and wines to be exported. As commercial wealth grows at a much faster rate than farm wealth and transport by sea is much cheaper than by land, Athens became a bustling commercial port city. As slave labor was the labor traditionally used for manufacturing, industry and transport, and as manufacturing, industry and trade dominated the Athenian economy, the situation in the feudal states was reversed: slave labor became the basic labor. Athens, in fact, arose on the backs of a virtual sea of slaves. Furthermore, the slave-owning capitalist class for a time became the dominant class. In addition to its rich trade in agricultural and manufactured products, Athens had great quarries and famed silver mines. This potential wealth turned into actual wealth by the extraordinary profits derived from slave labor is the main key to the historical phenomenon known as ancient Athens, providing the socio-economic base for its celebrated culture, its poets and playwrights, its historians and philosophers. Its political base lay in that uneasy balance of power between capitalist and landowning interests that produced democracy for these classes and hence considerable freedom of expression, artistic or otherwise, within their class peripheries. These freedoms, however, did not extend to women. Upper class women were largely deprived of property rights, excluded from the professions and secluded in the household. Both class struggle and female rebellion were continuous and sometimes ferocious, as the Athenian drama suffices to demonstrate. If these struggles appear in muted form in the works of the philosophers, they were nevertheless basic.[15]

As Plato (c. 427-347 B.C.), traditionally paired with Aristotle as the most famed of the Athenian philosophers, came from an aristocratic, landowning family, it is hardly surprising to find that his philosophy is akin to the (earlier) Hinduism of the landowning priesthood of India. Like them, he placed emphasis on changelessness. Plato's God, which he called the One, is a transcendental, creative force, indivisible, static and eternal, compounded of ideal "forms" of the Good, the True and the Beautiful. Unfortunately this God has to work with matter, seen as an inferior and inert substance; hence these perfect "forms" are distorted in their actual appearance on earth. The human mind is akin to the One, and existed in the One before birth, fully developed. On earth it is hindered in its attempts to blend again with the One by the human body, which, like matter, is viewed as an inferior – and evil – entity. It keeps the mind or soul – Plato did not distinguish between the two – from its natural affinity with the One:

> And were we not saying long ago that the soul when using the body as an instrument of perception . . . is then dragged by the body into the region of the changeable, and wanders and is confused; the world spins round her, and she is like a drunkard, when she touches change?

However, if one lives an ascetic life the mind may after death achieve union with the One and thus become immortal. As this asceticism did not extend to homosexual relations, it is apparent that it was primarily anti-feminist.[16]

Platonic doctrine is largely a sophisticated extension of Hinduism. Unlike Hinduism, however, it was not a mass religion but a ruling-class, male-oriented philosophy, comfortingly stressing the basic stability of the essence, and regarding change as evil. It preached an exclusive upper class, male immortality. Slaves, serfs and women could not acquire the heights of ascetic contemplation required for admission to the One. It implied – as did Hinduism – that non-aristocratic souls would transmigrate into the bodies of animals; and Plato's view of immortality appears as a philosophical projection of the male elitist ruling class state he depicted in *The Republic*. Plato expressed his philosophy in dialogue form, his main proponent in the dialogues being his teacher, Socrates, who would put forward a proposition to which other speakers would reply with an opposite proposition (thesis-antithesis) and the two would be pushed forward in opposing "dialectic" (*dialektikos:* debate) until a "synthesis" or conclusion was reached. True, the propositions of Socrates were usually of a simplistic either-or nature and if one naively accepted them the trap closed; but the method, with its dramatic interplay, had potential for philosophical and scientific probing.

Aristotle – born in an Athenian colony and hence not, like Plato, part of the Athenian aristocracy – attacked what he considered the non-rational and mystical elements in Plato's thought but was one with him in believing in an eternal and changeless force beyond matter: "Since there must always be motion without intermission there must necessarily be something eternal, a unity or it may be a plurality, that first imparts motion, and this first movement must be unmoved." Although Aristotle, unlike Plato (but like the earlier Rigveda poets and the Taoists), leaves the nature of his "Prime Mover" undefined, the basic concept is idealist, as is indicated by his argument that there are purposeful goals or "ends" inherent in nature. In coming to these conclusions Aristotle based his reasoning not on the analysis of facts but on abstract logic. For example: it is the nature of man to be intelligent, hence he has hands (which is apparently intended as a refutation of the materialist view of Anaxagoras that human intelligence arose from "the possession of hands"). The philosophical difference between Plato and Aristotle, then, is not basic, as is usually suggested, but is between two gradations of idealism. As he ushers Plato's mystical One out the front door, he lets his Prime Mover in the back door.

In his ethical and other views Aristotle, like Confucius, advocated "moderation" and "the middle way," combined with mass subordination: "It is clear then, that some men are by nature free, and others slaves, and that for these latter slavery is both expedient and right." In his general thinking Aristotle tended to favor a system of categories:

> Expressions which are in no way composite signify substance, quantity, quality, relation, place, time, position, state, action, or affection. To sketch my meaning roughly, examples of substance are "man" or "the horse"; of quantity, such terms as "two cubits long" or "three cubits long"; of quality, such attributes as "white," "grammatical." "Double," "half," "greater," fall under the category of relation; "in the market place," "in the Lyceum," under that of place; "yesterday," "last year," under that of time.

All this seems elementary enough but, in fact, it began a method of thinking that is both mechanical and remote from reality. Although Aristotle admits the existence of "contraries," he denies that they can overlap: "it is impossible for any one to believe the same thing to be and not to be, as some think Heraclitus says ... it is impossible that contrary attributes should belong at the same time to the same subject."

The "contraries," he states in another place, "are not affected by one another." (If so, development is impossible.) Applied to moral philosophy: "Nor can the same one action be good and bad." Aristotle, in short, worked out a system of static abstractions which he imposed on reality and saw as isolated absolutes. He has no sense, as did Heraclitus and other Greek materialists, of interactive opposites with an area of interblending between them. Sometimes, indeed, he simply spreads idealist confusion: "We must consider also in which of two ways the nature of the universe contains the good and the highest good." Clearly, the universe contains neither "good" nor "highest good," nor evil or any other moral quality.[17]

The early approaches to science in India were further developed by the Greek ruling classes, whose commercial economy needed technology. Archimedes, for instance, invented the compound pulley and the hydraulic screw and performed experiments in buoyancy, all of which required general scientific theory and mathematics. Strabo and other Greek thinkers – contrary to the usual view of "Greek thought" today – also used the experimental method. The philosophy reflecting these activities was most fully developed by Theophrastus, who attacked both Aristotle's Prime Mover and "ends" (teleological) theories. Why, he asked, as had the Jains before him, should we not assume that motion is an inherent part of matter? A question of basic philosophical import, for if it true then no external Prime Mover (or God) is necessary. As for the famed teleological doctrine of ends:

> With regard to the view that all things are for the sake of an end and nothing is in vain, the assignation of ends is in general not easy, as it is usually stated to be.

If, that is to say, one investigates the actual phenomena of nature and does not impose an abstract – basically theological – concept on them, the Aristotelian "goals" theory collapses under the weight of a virtually infinite complexity of events and causes.[18]

Prior to Theophrastus there had been a Greek materialist school whose best-known members were Heraclitus and Democritus (treated in Marx's doctoral thesis), and whose works have survived only in fragments. Democritus developed an atomic theory of matter, rejecting the "four elements" (earth, air, fire, water) view, an atomic theory perhaps based on that of the Jains but rejecting their mystical ("spirit") elaborations. Heraclitus, whose works now exist only in fragments declared that "becoming" (change) and not "being" (the static Platonic Godhead) was the essence of reality and that development took place by a conflict of interchangeable opposites in an eternal, non-created, world of fluid interactions:

> This world, which is the same for all, no one of gods or men has made; but it was ever, is now, and ever shall be an ever-living Fire, with measures of it kindling, and measures going out.

> We step and do not step into the same rivers; we are and are not.
> Homer was wrong in saying: "Would that strife might perish from among gods and men!" He did not see that he was praying for the destruction of the universe; for, if his prayer were heard, all things would pass away. . . .
> All things come into being and pass away through strife.[19]

These views of Heraclitus may have had some roots in the mixing-of-opposites views of the Taoists but they envisage the opposites as in conflict and not as circular. Furthermore, Heraclitus, like Democritus and unlike the Taoist skeptics is a materialist, viewing matter itself as the primal, creative entity. His views, like the experiments of Archimedes, represent the advancing commercial element in Greek slave-commercial civilization just as Plato's primarily represent those of the landed aristocracy. Thus the argument on whether "being" or "becoming" was basic is not purely philosophical, as it appears on the surface to be, but, however indirectly, reflects the class struggle between landowning and commercial interests, the former desiring stability, the latter advance.

In recent years as we have learned more about the extensive trade routes that early arose between slave-commercial Europe and feudal Asia, it has become apparent that there were cultural exchanges between them as well as between West and East Asia. The parallels between Plato and Hinduism and between the Greek materialists and the Jains can hardly be coincidental, nor can the fact that Aristotle's "middle way" ethical and antimystical approaches were anticipated by Gautama and Confucius. Nor that "Tao" was seen as a "prime mover." None of this, of course, was a simple matter of copying but of seed influences from basically similar societies being further developed by the sharper socio-economic dynamics of slave-commercialism.

As the Athenian empire was declining, the Italian was rising, and by the first century B.C., a state had developed in Roman Italy similar to that of Athens, one based on slave labor, with a rough balance of political power between commercial and landowning interests. Once again, considerable freedom of expression existed for upper class male intellectuals.

One of the leading political figures and writers connected with

Roman commercial interests was Marcus Tullius Cicero, a "new man" – i.e., not an aristocrat – who expressed not only political but philosophical views, most notably in his skeptical *Dialogue on the Gods,* that ran counter in some ways to the aristocratic interest. Apparently connected with Cicero's intellectual circle was the poet Lucretius (99–55 B.C.) whose philosophical epic, *On the Nature of Things,* is the only full exposition of materialism that has survived from the Greek and Roman slave-commercial civilizations.

When we read *On the Nature of Things,* we begin to see the scope of his Greek materialist predecessors. Read only in the surviving fragments – their full works were apparently destroyed by their religious opponents – their sophistication and scope are blurred. Lucretius, however, as Alban D. Winspear has shown, was not simply repeating the formulations of his Greek predecessors but developing – in, as Marx notes, marvelous verse – the essence of their philosophy. *On the Nature of Things* in time became the main intellectual wellspring for European materialism. It provided the base some 1,800 years later for the French 18th century materialists, and was a major influence in the philosophical development of Marx. Its (general) anticipations of later scientific concepts are startling. As are its dialectical insights.[20]

Lucretius followed Democritus in declaring that nothing exists but "atoms and the void," atoms including "particles of light" from the sun and the void being not only space in general but also space between atoms: "There are clear indications that things that pass for solid are in fact porous . . . Noises pass through walls." All the complex phenomena of nature and life were the result of the "compounding" of atoms: "Material objects are of two kinds, atoms and compounds of atoms." The compounds arose from the "hooking" together of atoms:

> Is not copper soldered to copper by nothing but tin . . . When the textures of two substances are mutually contrary, so that hollows in the one correspond to projections in the other and vice versa, then connexion between them is most perfect. It is even possible for some things to be coupled together, as though interlinked by hooks and eyes.

There is no God and no creation: "Nothing can ever be created by divine power out of nothing." The universe is infinite and has existed from eternity: "the sum total of the universe is everlasting, having no space outside it." Its present state is the result of chance atomic combinations: "multitudinous atoms swept along in multitudinous courses through infinite time by mutual clashes and their own weight, have come together in every possible way and realized everything that could be formed by their combinations." At first, there was nothing but "a hurricane raging in a newly congregated mass of atoms of every sort. From their disharmony sprang conflict." From this conflict arose "the stars that crown the far-flung firmament" and the "sun's bright disk" – and the earth.[21]

The atoms themselves, although varied in nature, are indestructible:

> But, since I have already shown that nothing can be created out of nothing nor any existing thing be summoned back to nothing, the atoms must be made of imperishable stuff into which everything can be resolved in the end, so that there may be a stock of matter for building the world anew.

Natural phenomena, however, although composed of unchangeable atoms, themselves change. The new arises not by divine creativity but from combinations and patterns of the old; and the world as we perceive it is different from the underlying reality:

> So you may realize what a difference it makes in what combinations and positions the same elements occur, and what motions they mutually pass on and take over. You will thus avoid the mistake of conceiving as permanent properties of the atoms the qualities that are seen floating on the surface of things, coming into being from time to time and as suddenly perishing. Obviously it makes a great difference in these verses of mine in what context and order the letters are arranged. If not all, at least the greater part is alike. But differences in their position distinguish word from word. Just so with actual objects: when there is a change in the combination, motion, order, position or shapes of the component matter, there must be a corresponding change in the object composed.[22]

Life had been produced by the Earth; first plant, then animal life, some surviving, some dying, by a kind of natural selection: "In those days, again, many species must have died out altogether and failed to reproduce their kind. Every species that you now see drawing the breath of life has been protected and preserved from the beginning of the world either by cunning or by prowess or by speed." There is no absolute dichotomy between mind and body. They interact through a "vital spirit" which is also atomic in nature, part, that is to say, of matter: "whatever is seen to be sentient is nevertheless composed of atoms that are insentient." Perception is the result of external, atomic stimuli: "all invisible objects emit a perceptual stream and shower of particles that strike upon the eyes and provoke sight." The mind, being part of the body, dies with the body. Plato's argument that the mind has ideas before birth is given short shrift: "If the spirit is by nature immortal and is slipped into the body at birth, why do we retain no memory of an earlier existence?" Lucretius goes beyond the dualism of the Jains on mind and matter into a consistent materialist view and considers body and mind, movement and sensation all as expressions of the same material force:

> The same reasoning proves that mind and spirit are both composed of matter. We see them propelling the limbs, rousing the body from sleep, changing the expression of the face and guiding and steering the whole man — activities that all clearly involve touch, as touch in turn involves matter. How then can we deny their material nature?

There is no immortality, no heaven and no hell; and no need to fear death:

> The old is always thrust aside to make way for the new, and one thing must be built out of the wreck of another. There is no murky pit of Hell awaiting anyone. There is need of matter, so that later generations may arise; when they have lived out their span, they will all follow you. Bygone generations have taken your road, and those to come will take it no less. So one thing will never cease to spring from another. To none is life given in freehold; to all on lease. Look back at the eternity that passed before we were born, and mark how utterly it counts to us as nothing. This is a mirror that Nature holds up to us, in which we may see the time that shall be after we are dead. Is there anything terrifying in the sight — anything depressing — anything that is not more restful than the soundest sleep?[23]

As we read *On the Nature of Things,* it becomes apparent that the materialists of the Greek and Roman slave-commercial civilizations had arrived at general solutions to philosophical and scientific problems that are borne out by modern science and that the intellectual roots of dialectical materialism stretch back further than is generally realized. True, atoms have turned out to be not the solid objects that Lucretius thought, but his contention that all the phenomena of nature arise from their "compounds" anticipates the discoveries of atomic and molecular combinations – "hooks and eyes." His extension of this theory in the argument that all these phenomena arose from chance combinations on a virtually infinite scale anticipates modern scientific views on the mechanisms behind both cosmic and biological evolution. And he anticipates, again in a general way, the "natural selection" view of evolution. His concept of cosmic "conflict" arising from atomic "disharmony" embodies the essence of contradiction. His view of the new arising from the "combinations and positions" of the old – with roots going back to the Jains – contains the essence of the dialectical materialist view that the new (quality) arises solely from additions and combinations of the old. So too, as with Democritus, of the concept of a basic "conflict" inherent in nature. And that of mind arising from matter.

On the other hand, Lucretius was not as consistently scientific as were some of his Greek predecessors, as may be seen in such arguments as that the sun "cannot be either much larger or smaller than it appears to our senses," and, the mind is "firmly lodged in the mid-region of the breast." Unlike Theophrastus or Strabo, he relied primarily on reason or "common sense" and not on scientific demonstrations. He did not appear to know that the Earth is round – as the Greek scholar Eratosthenes had ingeniously proved – or that Aristarchus of Samoa had demonstrated that the sun was larger than the earth or that Alomaeon of Croton, discovering the optic nerve, had declared the brain to be the organ of the mind.[24]

If, however, we put together the views of Lucretius and the Greek materialists, the wonder is not how little they knew but how much understanding of the general nature of reality they were able to derive from that little. If we ask ourselves why they were able to do this, the social reason must lie in the demand (however limited) for technology by slave-commercial capitalism, which in turn led to certain general scientific truths even in a science without telescope or microscope. The philosophical answer must lie in the fact that what we perceive with our senses is part of reality and not a separate entity from that revealed by science. Hence, reasoning by analogy can often give a general approximation of the actual material process at work. This reasoning was allowed open expression because of the close balance of landowning and commercial interests in Athens and Rome. We might note, too, that the Greek and Roman materialists were denegrated by later feudal and bourgeois thinkers whose strong, if sometimes indirect, religious bias moved them to elevate the idealists Plato and Aristotle to virtual godhead status.

The philosophical views we have just surveyed are those of the upper classes of early feudal and slave-commercial societies. Although in the latter (Athens and Roman Italy) the commercial component was relatively greater than in the feudal (Indian, Chinese) and the views expressed consequently more advanced both were on the same general ideological level. Both feudal and slave-commercial societies were based on the exploitation of unpaid labor and the socio-economic structures of both were essentially the same, namely that of landowners and capitalists at the top, living in luxury, and peasant-slave masses at the bottom, living in poverty. Politically the situation was different in that in feudal societies the landowners formed a ruling-class oligarchy and the capitalists a fitful opposition, whereas in slave-commercial societies there was some sharing of power between capitalist and feudal interests with periods of capitalist dominance. As economic and political control in both societies was held by men, the societies were male-dominant, with ruling class women without effective intellectual expression and barred from political power and lower class women bearing the double burden of class and sexist exploitation. Hence, these philosophical views were not only those of the upper classes but of the men of those classes.

In considering the views of these societies, we have noted two trends, emanating respectively from the landowning and commercial classes. These views, however, were not usually formulated by the landowners or capitalists themselves but by intellectuals connected with them, whose views, arising in the vortex of social struggle, often reflected mixtures of both. Thus although Plato represented a consistent aristocratic position, politically and philosophically, Aristotle, the great compromiser, expressed views that reflected partly landowning and partly commercial interests which, of course, on some matters were identical. This was also true of Sumerian, Indian, Chinese and other feudal civilizations. Philosophers representing primarily the feudal class viewpoint were usually idealist and religious, preaching doctrines of changelessness, immortality and asceticism. As Hinduism reveals, the doctrine of immortality was intended both to narcotize the masses and provide the upper classes with moral justification for war or oppression. Asceticism implied justification for the suppression of women. On the other hand, views based, directly or indirectly, on commercial interests tended towards skepticism (Gautama, Confucius, Aristotle and Cicero) and sometimes toward materialism (Mahavira, Democritus, Heraclitus, Epicurus, Theophrastus and Lucretius).

Within these common patterns there were degrees of development. Slave-commercial civilizations produced greater concentrations of wealth than did feudal civilizations, a more powerful capitalist class and consequently both political flexibility and greater technological and scientific advance. These qualities laid the base for an assertion of human power in defiance of nihilistic fatalism, for instance, in Sophocles' hymn to humanity ("wonderful in skill," with "thought as swift as wind"):

Many the forms of life,
Wondrous and strange to see,

But nought than man appears
More wondrous and more strange.

Plato's idealism was more logically structured than Hinduism, Aristotle's moderation and anti-mystical theories were more systematically developed than those of Buddha or Confucius, the materialism of Heraclitus and Lucretius was more consistent and extensive than that of Mahavira and the Jains. But both feudal and slave-commercial civilizations put the same general ceiling on philosophical (and social) thought, namely that determined by their mass-exploitive foundations. Neither had a concept of future historical progress and could not, except fragmentarily, apply their materialist and dialectical views to society. And their idealism, unlike Hegel's, was static.[25]

Commercial Capitalism: From Bacon to Hegel

What is generally known as the "decline of the Roman Empire" was actually a decline in Roman commercial capitalism, a decline (underway by 300 A.D.) that resulted primarily from the limitations placed on technological advance by slave labor. (Why advance technology when there were slaves to do everything?) It was speeded up by attacks from other peoples, especially those who had felt the lash of Roman slavery. As a result, Roman Italy collapsed back into the feudal state from which it had originally emerged. (The slaves were "hutted," that is to say, were put in peasant family units.) In the rest of Europe, society was either in a low feudal state or at a pre-feudal farming level. By about the year 1000, feudal states existed in most of Europe. Bourgeois historians have obscured the fact that this feudalism was essentially the same as what had existed in Asia for many centuries. The differences were, in fact, minor. The lot of the "sturdy" British peasant and that of the Chinese or Indian serf did not differ materially; nor did the bloodsucking brutality of their exploiters. By about 1300, however, European feudalism in areas with large sea-access began to acquire an unusually strong commercial component, as shown in the development of such port cities as Genoa and London, the great trade routes of the Mediterranean, and the European northern seas trading – especially in wool, lumber and leather – of the Hanseatic League. By 1600, some of these port cities had become virtual city-states. By 1700, the commercial bourgeoisie had seized political power in two European nations, Britain and Holland, colonization and intercontinental trade interactively expanded.

With the breakdown of Roman slave-commercial society and the rise of feudalism in Europe, culture, including philosophy, naturally dropped to the feudal level. Materialism and skepticism declined and theology rose, a theology embodied in the Nicene Creed (written in 325 and revised in 362), which may be summarized as follows:

(1) The universe and all it contains was created by a spiritual substance called God ("the maker of all things, visible and invisible").

(2) This God is of a triune nature – Father, Son, and Holy Ghost (Spirit).

(3) By divine intervention in the process of nature one of the aspects of this "Trinity" entered the uterus of a carpenter's wife in Palestine about 4 B.C. ("was incarnate of the Holy Ghost and the Virgin Mary").

(4) This aspect, receiving bodily form as Jesus, elected to be executed in order that those who believed in his divine nature might receive immortality ("whose kingdom shall have no end").[26]

Although differing in specifics, these mystical concepts are fundamentally similar to those which had arisen centuries earlier in Asian feudal states. The earliest of the religions in the direct line of evolution into the Judaic and then the Christian was the Persian Zoroastrianism, which posited a God, his prophet (Zoroaster – born of a "virgin birth"), sacred works that embodied the "word" of God, a Satan (Ahriman), a "last judgment," and immortality – in a heaven or hell. This strain of religion differed from Hinduism, with its "transmigration" immortality, in its heaven – hell views, but it shared with it a belief in God, immortality, divine rule and judgment; and some of these ideas went back into early farming and hunter-gathering societies. As we read the "creeds" of these various religions, it becomes clear that they are not reasoned conclusions but like their predecessors class-rule fantasies exploiting human fears:

I, stranger and afraid
In a world I never made.

Certain theologians, however, both Christian and Mohammedan – which also ultimately stemmed back to Zoroastrianism – began a reasoned defense of their views. Among the Christians, one of the most influential theologians was Thomas Aquinas (1225?-1274) who added intellectual embellishments largely derived from Aristotle and Plato.

His famed "proof" of the existence of God, for instance, is simply a repetition of Aristotle's Prime Mover and predestined "ends" arguments:

> The first and more manifest way is the argument from motion. It is certain, and evident to our senses, that in the world some things are in motion. Now whatever is in motion is put in motion by another, for nothing can be in motion except it is in potentiality to that towards which it is in motion.

Aquinas' method is not one of reasoning from facts but of juggling abstractions. In this regard he did not advance beyond such earlier Asian thinkers as the Hinduist theologians. Aquinas also argued that "revelation" was a source for "truth," thus adding a dollop of mysticism to his ultra-abstractionism.[27]

Although there were counter-currents in the church, as shown in Roger Bacon's emphasis in the 13th century on "the principles of this science which is called experiments," no major philosophical advance was possible until commercial capitalism began to open cracks in the feudal monolith. As this time commercial development came not on a slave-capitalist but a paid-labor-capitalist basis, it was able to advance technology (partly by developing that already existing in the commercial enclaves of China: printing, gunpowder, the compass); and technology needed science (Copernicus, Kepler, Galileo).[28]

The first philosopher of this new movement, Francis Bacon (1561–1626), was born into an England with unusual commercial development, based mainly on its agricultural products, especially wool and lumber, and its imperialist expansion into the "New World" of the Americas, so vividly described by Marx in *Capital.* Bacon was both a political (Lord Chancellor) and a philosophical leader. Among his contemporaries were such colonial explorers as Sir Francis Drake and Sir Walter Raleigh; such soldier intellectuals as the Earl of Essex and Sir Philip Sidney (whose family owned a cannon factory), as well as the merchants and bankers of the city of London, whose private capital built and maintained the great naval fleet that destroyed the Spanish armada. In literature, the historical movement was reflected in Shakespeare – most directly in *Henry V;* in Edmund Spenser, with his epic *Fairy Queen* dreams of British imperial glory; Marlowe with his religious skepticism and exaltation of human achievement (*Tamerlane, Doctor Faustus*); Raleigh with his ambitious *History of the World;* and Sidney, countering the feudal rhetoricians in his *Defense of Poesie.* It was also an era in Britain in which science was beginning to take root and the feudal church had been weakened. Gabriel Harvey, the discoverer of the heart's function as a pump and of the circulation through it of the blood, was also a contemporary of Bacon's and his work pointed the way to Newton, Boyle and the Royal Society of London for Improving Natural Knowledge. Of them all, Bacon was the most advanced thinker, representing the amalgamation of aristocratic-commercial interests (Essex, Raleigh, Sidney) with those of the London manufacturers, merchants and bankers, the former group giving the movement (as in Italy) an air of confidence and magnificence, the latter keeping its feet on the ground.

The boldness of Bacon's thought is immediately obvious in his frontal attack on Plato and Aristotle: "we have as yet no natural philosophy that is pure; all is tainted and corrupted; in Aristotle's school by logic; in Plato's by natural theology." By Aristotle's "logic," Bacon meant Aristotle's "deductive" method – establishing abstract categories without a systematic examination of the available facts and then fitting reality into them.

> There are [Bacon contended] and can be only two ways of searching into and discovering truth. The one [Aristotle's] flies from the senses and particulars to the most general axioms, and from these principles, the truth of which it takes for settled and immovable, proceeds to judgment and to the discovery of middle axioms. And this way is now in fashion. The other derives axioms from the senses and particulars, rising by a gradual and unbroken ascent, so that it arrives at the most general axioms last of all. This is the true way, but as yet untried.[29]

The essence of the new philosophy was to be science, research and experience: "For when philosophy is severed from its roots in experience, whence it first sprouted and grew, it becomes a dead thing." The "particulars" must be examined "in an orderly and systematic way." Such philosophies as Plato's, Aristotle's and, he implies, that of the church of his own day, not only spread debilitating pessimism but fostered economic retrogression:

> The effect and intention of these arguments is to convince men that nothing really great, nothing by which nature can be commanded and subdued, is to be expected from human art and human labour. Such teachings, if they be justly appraised, will be found to tend to nothing less than a wicked effort to curtail human power over nature and to produce a deliberate and artificial despair. This despair in its turn confounds the promptings of hope, cuts the springs and sinews of industry, and makes men unwilling to put anything to the hazard of trial.[30]

Bacon here not only – as in the comment on Plato – attacks theology as obfuscation but suggests its inherently reactionary character, its doctrines denigrating humanity's efforts to advance (by proclaiming the absoluteness of God's power). Moreover the kind of advance Bacon advocates – to build up the "sinews of industry" – illustrates his pro-commercial and anti-feudal emphasis.

In considering the Greek myth found (among other places) in Plato's *Phaedrus,* that Love was the creative force behind the universe, Bacon gave it his own interpretation:

> The fable relates to the cradle and infancy of nature, and pierces deep. This Love I understand to be the appetite or instinct of primal matter; or to speak more plainly, *the natural motion of the atom;* which is indeed the original and unique force that constitutes and fashions all things out of matter.

Thus, like Lucretius, Bacon – in spite of some perfunctory references to God required at the time – was, as Marx noted, a materialist, seeing the fire of the universe and life alike as residing in atomic activity. Like Theophrastus and other Greek materialists, he believed motion to be of the essence of matter, scornfully dubbing Aristotle's and Aquinas' separation of motion from matter "blind and babbling." Matter was both eternal and eternally changing: "Natural bodies refuse to be destroyed, to be annihilated; rather than this they will turn themselves into something else."[31]

With Bacon riding the confident sweep of the new capitalism, the philosophy of science, previously only implied, becomes a reality.

In France where neither capitalism nor science had developed as they had in Britain, René Déscartes (1596-1650) produced, a few decades later than Bacon, a betwixt-and-between philosophy. Although he commended science and condemned dogma, his basic philosophical position was one of egocentric idealism:

> It was absolutely essential that the "I" who thought this should be somewhat, and remarking that this truth "I think, therefore I am" was so certain and so assured that all the most extravagant suppositions brought forward by the skeptics were incapable of shaking it, I came to the conclusion that I could receive it without scruple as the first principle of the Philosophy for which I was seeking.

Here in Déscartes we can see the beginning of what was to become a new central base for idealism, namely the passive, observing "mind." In Déscartes this mind – or "soul" – is seen as a unique essence:

so that this "me," that is to say, the soul by which I am what I am, is entirely distinct from body, and is ever more easy to know than is the latter; and even if body were not, the soul would not cease to be what it is.

Like Plato, then, Déscartes sees an absolute dichotomy between mind and body. The body is a simple mechanism unconnected with the soul or mind, the same in people as in animals (which he regarded as mindless automatons). And above both mind and body stands God, seen not, however, as in Plato, as a mind substance but as something unique. Unlike Bacon, Déscartes also continued Aristotle's method of first constructing abstractions and then fitting facts into them.[32]

Following Déscartes among the Continental philosophers, Baruch Spinoza (1632-77), the son of a Jewish merchant in the commercial Netherlands, advanced views – "Nature herself is the power of God under another name" – that led to anti-theological positions including a repudiation of miracle and prophecy. His anti-feudal base is shown in his attack on dictatorial governments that use "the specious garb of religion so that men may fight as bravely for slavery as for safety, and count it not shame but highest honour to risk their lives for the vainglory of a tyrant." Attacked as an atheist, Spinoza (truthfully) denied the charge but his views laid a base for the skeptical deism of Voltaire and Paine.[33]

The rapid development of paid-labor commercial capitalism in Britain brought about a new level of scientific advance, and this presented a new set of philosophical problems. Newton – anticipated by Lucretius – argued that light was the result of the motion of microscopic particles. It could be broken down by a prism into primary colors. Were these colors in the light or in the mind? Could they be produced in the mind by the action of particles that themselves had no color? Examining this and similar problems, John Locke (1632-1704), who once had aspired to chemistry and medicine, declared that certain qualities, which he called "primary" – "solidity, extension, figure and mobility" – actually existed in nature but that other qualities ("secondary") – "colours, sounds, tastes" – do not; they are, in fact, "nothing in the objects themselves but powers to produce various sensations in us by their primary qualities."

Locke also attacked the Platonic concept that the mind had "innate ideas" (which Lucretius also had rejected). The mind at birth was "a blank tablet"; all that it later contained came from the senses through experience as "complex ideas" are "compounded" from the "simple ideas which we receive from sensation and reflection." In putting forward such views Locke was attempting to undermine the primarily feudal (Catholic) theological doctrines of revelation and dogma. But he was no atheist: "those are not all to be tolerated who deny the being of God." And although he attacked aristocratic privilege he believed in private property: "The great and chief end therefore of men's uniting into commonwealths and putting themselves under government is the preservation of their property." In short, Locke, a Commissioner of the Board of Trade, produced an anti-feudal, common-sense bourgeois philosophy. In doing so, he helped to build new foundations for materialism although he was not himself a materialist but a dualistic idealist, seeing mind and matter as separate entities under God.[34]

Locke's brand of idealism, however, sounded suspect to the theologians, for it opened the door to science, and one of their number, Bishop George Berkeley (1685-1753), led the philosophical attack on him, developing as he did so, his own brand of idealism. If, the Bishop contended, mind and matter are, as Locke implies, essentially different entities, it is impossible that they could interact. Inert matter could not make an impact on spiritual mind. Therefore what appears to be matter must itself be a mind substance:

> When in Broad daylight I open my eyes, it is not in my power to choose whether I shall see or not, or to determine what particular objects shall present themselves to my view: and so likewise as to the hearing and other senses; the ideas imprinted on them are not creatures of my will. There is therefore some other Will or Spirit that produces them.

> But after all, you say, it sounds very harsh to say we eat and drink ideas, and are clothed with ideas. I acknowledge it does so—the word *idea* not being used in common discourse to signify the several combinations of sensible qualities which are called *things*. [35]

Berkeley thus went beyond the objective idealism of Plato – which posited the existence of both an external world of matter and a God (the One) – and argued that the world only appeared to be material; in reality it was a spiritual substance, a projection of the Mind of God. There was nothing but this Mind and the human mind, which was akin to it. Something of this view had, of course, existed previously, for instance in the mystical aspects of Taoism or Plato's view of the relation of the human and divine mind. Its subjectivism had roots in Déscartes, its view of God as Nature in Spinoza. But Berkeley turned the whole into a new subjectivist-idealist system, which has affected philosophy into the 20th century.

David Hume (1711-76), a Scottish middle class philosopher and historian, performed the same systematizing services for skepticism, and in doing so became one of Berkeley's (and indeed, all idealist philosophy's) most devastating critics. How, Hume asked, does Berkeley know that his "Will or Spirit" exists if he knows nothing but his own sensations? Could not what he calls God simply be one of these sensations? Similarly how does Locke know that there are "primary qualities" if all we are aware of in the mind are the "secondary qualities"? There is, in fact, no actual evidence that our sensations or thoughts are stimulated by a force beyond them, either material or spiritual.

> The fundamental principle of that philosophy [Locke's] is the opinion concerning colours, sounds, tastes, smells, heat and cold; which it asserts to be nothing but impressions in the mind, deriv'd from the operation of external objects, and without any resemblance to the qualities of the objects . . . I believe many objections might be made to this system: But at present I shall confine myself to one, which is in my opinion very decisive. I assert, that instead of explaining the operations of external objects by its means, we utterly annihilate all these objects, and reduce ourselves to the opinions of the most extravagant scepticism concerning them. If colours, sounds, tastes, and smells be merely perceptions, nothing we can conceive is possest of a real, continu'd, and independent existence; not even motion, extension and solidity, which are the primary qualities chiefly insisted on.[36]

One of Hume's main motives in putting forward these skeptical arguments was anti-religious. If we know nothing but sensations, we cannot know whether or not there is a God. If we assume that a spiritual being created matter, we could also assume a creator for the spiritual being, and so on in a ridiculous "infinite regress" – a God who created a God who created a God: "If the material world rests upon a similar ideal world, this ideal world must rest upon some other; and so on, without end. It were better, therefore, never to look beyond the present material world." Nor can we argue, as did Aristotle, for a First Cause (God) because causation is not inherent in nature. All we actually observe is a succession of sensations or thoughts; it is we who attribute "cause" to the succession:

> The first time a man saw the communication of motion by impulse, as by the shock of two billiard balls, he could not pronounce that the one event was connected: but only that it was conjoined with the other. After he has observed several instances of this nature, he then pronounces them to

be connected. What alteration has happened to give rise to this new idea of connexion? Nothing but that he now feels these events to be connected in his imagination, and can readily foretell the existence of one from the appearance of the other.[37]

Hume thus – not without a certain wry humor – reduced everything (including space and time) to a stream of internal sensation with "cause" being nothing but meaningless succession. The idealist baby was thrown out with the philosophical bathwater, by the application of pure logic to pure abstraction. Skepticism had turned in upon itself like a snake eating its tail.

The ideas of Locke, Berkeley and Hume were the product of a comparative political calm in Britain. The British anti-feudal revolution led by Oliver Cromwell in the 1640's had ended in a compromise government, balancing greatlandowning and business interests. In France, however, with its greater locked-in landmass, although commercial capitalism had expanded, the bourgeois feudal aristocracy still ruled supreme; business and professional men were excessively taxed and repressed. The law was the law of the manor and the church; parliamentary government did not exist. When Voltaire offended a great aristocrat he was beaten like a dog in the street and had no recourse in the law. Thus when the French bourgeois intellectuals – often, like Voltaire and Rousseau, driven into exile – began to form a philosophy it was violently anti-feudal and anti-clerical, much different in tone and content from the studied British expositions. Denis Diderot (1713-84), who, like Voltaire, was imprisoned and constantly persecuted for his ideas, and Paul Henri Holbach (1723-89) went beyond the dualistic idealism of Locke and the nihilistic skepticism of Hume into materialism.

Diderot and Holbach shared Voltaire's hatred of established religion. "Religion," wrote Holbach, "is the art of intoxicating people with religious fervor, to prevent them being cognizant of their troubles, heaped upon them by those who governed." But unlike Voltaire, a "deist" who left the door open for a vague deity, they returned to Lucretius's concept of the universe as eternal and matter as the only existing entity: "The universe, that vast assemblage of everything that exists, presents only matter in motion." There was no need of a Prime Mover: "The essence of matter is to act." Within matter there are immutable laws (revealed by Kepler, Newton and others) which act in certain ways of "necessity": "This irresistible power, this universal necessity, this general energy, is then only a consequence of the nature of things, by virtue of which everything acts, without intermission of constant and immutable laws." Both body and mind are simply manifestations of matter, the result of certain natural combinations: "In short, experiment proves beyond a doubt that matter, which is as inert and dead, assumes sensible action, intelligence and life, when it is combined of particular modes." In regard to the arguments on the reality of the external world advanced by Berkeley and Hume, Holbach answered (as Bacon would have): "The only test of truth is experience."[38]

Engels' later charge that the French materialists were "mechanical," unable "to comprehend the universe as a process," is only partly true; as witness the following in Diderot:

> Everything changes and everything passes away, only the whole endures. The world is for ever beginning and ending; each instant is its first and its last; it never has had, it never will have, other beginning or end. In this vast ocean of matter, not one molecule is like another, no molecule is for one moment like itself.

Diderot clearly had a concept of "the universe as a process" (and not just of separate "mechanical" entities).

He also had a concept of developmental change. In fact, building on the science of the day, he was able to go beyond Lucretius in his evolutionary views: "The vegetable kingdom might well be and have been the first source of the animal kingdom, and have had its own source in the mineral kingdom; and the latter have originated from universal heterogeneous matter." Biological matter, then, developed from physical matter. And once biological matter began, it evolved into new forms by, as hinted by Lucretius, a kind of natural selection: " . . . that monsters annihilated one another in succession; that all the defective combinations of matter have disappeared, and that there have only survived those in which the organization did not involve any important contradiction, and which could subsist by themselves and perpetuate themselves." But that – again like Lucretius – Diderot had no solid, scientific basis for his evolutionary theory is shown by such conjectures as the following: "The elephant, that huge organized mass a sudden product of fermentation! Why not?" And both Diderot and Holbach had no explanation for the origin of Man, no concept of human evolution: "There have perhaps been men upon earth since eternity." Following Lucretius they regarded mind as a manifestation of matter, and Holbach speculated that it might be the result of invisible particles similar to fermentation. Diderot, however, going beyond Lucretius, argued that the mind had arisen not simply from "atoms" but from biological matter as the result of "heat and motion":

Do you see this egg? With this you can overthrow all the schools of theology, all the churches of the earth. What is this egg? An unperceiving mass, before the germ is introduced into it; and after the germ is introduced, what is it then? still only an unperceiving mass for this germ itself is only a crude inert fluid. How will this mass develop into a different organization, to sensitiveness, to life? By means of heat. And what will produce the heat? Motion . . . If you admit that between the animal and yourself the difference is merely one of organization, you will be honest; but from this there will be drawn the conclusion that refutes you; namely that, from inert matter, organized in a certain way, and impregnated with other inert matter, and given heat and motion, there results the faculty of sensation, life, memory, consciousness, passion and thought.

Today when we can begin to see the actual evolutionary stages in which the human mind developed from living matter and living matter from matter, we can only wonder at Diderot's insights in an age when biological research had barely begun, and the skill with which he counters the theological argument of the spirituality of mind. Whatever its specific nature, however, Diderot and Holbach agreed with Lucretius that the mind dies with the body, on which it is dependent. To argue that the mind or soul is immortal is like pretending "that a clock shivered into a thousand pieces, will continue to strike the hour."[39]

Following Diderot and Holbach, evolutionary theories continued to develop: "All vegetables and animals now existing were originally derived from the smallest microscopic ones, formed by spontaneous vitality." So wrote Erasmus Darwin (1731-1802), grandfather of Charles (1809-1882), who seems to have overlooked his significance. Erasmus Darwin argued further that "all warm-blooded animals have arisen from one living filament," and that "mankind arose from one family of monkeys on the banks of the Mediterranean." Geological discoveries began to produce concepts of geological time and development. These advances, along with those in chemistry and physics, especially experiments with electricity and magnetism, produced a new concept of matter, an essentially dialectical concept, as witness the following from Percy Bysshe Shelley's *A Refutation of Deism* (1813):

Matter, such as we behold it, is not inert. It is infinitely active and subtile. Light, electricity and magnetism are fluids not surpassed by thought itself in tenuity and activity; like thought they are sometimes the cause and sometimes the effect of motion.[40]

By Shelley's time, as a result of the experiments of John Dalton, Sir Humphrey Davy, Allesandro Volta and Luigi Galvani, the concept of matter was beginning to change from the solidity of Locke's "primary qualities" into a more "fluid" one. Electricity, magnetism and light were also forms of matter; so, too, the endless circling of atoms:

> *and they whirl*
> *Over each other with a thousand motions,*
> *Upon a thousand sightless axles spinning.*

Lightning was as much matter as was a rock. Shelley – in an age of revolution, industrial and social – also saw conflict at the heart of development in nature and society: "The laws of attraction and repulsion, desire and aversion, suffice to account for every phenomenon of the moral and physical world." Or in another formulation: "The laws of motion and the properties of matter suffice to account for every phenomenon, or combination of phenomena exhibited in the universe." In Shelley we can see, as it were, dialectical materialism in embryo.[41]

Following French materialism, the next major philosophical development came in Germany in the late 18th and early 19th centuries. At the time there was no unified German nation but a series of feudal states in which capitalist economic forces with their accompanying bourgeois political surge were beginning to grow. In repressive states, new views were often expressed more in cultural than in directly political form, as in Fichte's impassioned bourgeois nationalism, in the humanist aspirations of Goethe's *Faust,* the anti-feudal rebellion of Schiller's *Robbers,* the soaring symphonies of Beethoven, and to some degree in Kant and Hegel. When we begin to read the philosophical works of Kant and Hegel we soon find – in contrast to Bacon, Locke, Berkeley and Hume or the French materialists and skeptics – that they are difficult to understand. This is not because they are more profound but because they still followed the old methods of feudal theology, piling abstraction upon abstraction, sometimes within a framework of "categories." Coming from feudal states with capitalist stirrings, they are a curious mixture of the old and the new. Kant, for instance, although abjuring scientific method in his philosophical writings, was keenly interested in science, especially astronomy, early arguing that certain "stars" were really far distant nebulae showing up merely as points of light and that the solar system had evolved from swirling masses of gas – thus delivering the first major blow to the serene, unchangeable universe envisaged by Newton. In philosophy he was intent upon undoing the damage wrought by Hume's iconoclastic sweep. Hume, he apparently felt, had been able to wreak such havoc because of the particular idealist position taken by Berkeley. He is at some pains to distance himself from this position:

> The essence of idealism is the assertion that only thinking beings exist, and that the other things which we think we perceive in contemplation are merely concepts within thinking beings, concepts to which no object outside them corresponds. On the contrary, I say: things are given to us as objects of our senses existing outside us, but we know nothing about what they really are, we only know their appearances, i.e., the images they generate in us by affecting our senses. Therefore, I naturally recognize that there exist objects outside us, i.e., things about whose essence we know absolutely nothing but which we know through the concepts we receive as results of their impact on our senses and call bodies, a name thus denoting only the appearance of the object which is unknown to us but nevertheless real. How can one call this idealism? It is a complete antithesis of the latter.[42]

Although Kant here repudiates "idealism," actually the only idealism he is repudiating is the subjective idealism of Berkeley, of which he gives a rather garbled account. He is not repudiating the European mainstream of idealism stemming from Plato. Plato believed in the existence of both mind (divine and human) and "objects outside us." Like Plato, Kant also believed in *a priori* ideas, ideas innate in the mind at birth, and although, unlike Plato, he does not specifically designate God as the source of these ideas, this could certainly be implied. (Where else could they come from?)

All we know, Kant is arguing, are mental frameworks that are somehow built into the mind and by means of which it attempts to interpret, however vainly, the "objects of our senses." These frameworks are the famed Kantian "categories." Their basic – indeed, all-embracing – nature is shown by the fact that among them are time, space and causality. Time and space do not have external existence but are simply ways in which the mind regards reality. The categories, in effect, create a barrier between the mind and reality, for it is because the mind can know only the categories that it can never really grasp reality. There will always be a large unknowable area ("the thing in itself"). The categories, because they are viewed as rigid, as in Aristotle, result in insoluble contradictions: the world is both finite and infinite; God both exists and does not exist. Belief in God, however, although unprovable by reason was, Kant argued, justified by "faith." Kant, in short, attempted to shore up what was left of subjective idealism after Hume's assault by a retreat to faith and a scaffolding of innate "categories." Nor does his famed "starry heavens above me and the moral law within me" change the picture. As the "moral law" is perceived not as a socio-psychological entity but as an abstract *a priori* absolute, it (like the categories) could only emanate from God. And the "starry heavens" are God-directed. None of this, of course, is based on a single demonstrable fact but is solely the product of abstractionist assertion. Kant, in spite of his disclaimers is, as some of his contemporaries apparently charged, an idealist. The notion in some Soviet and Marxist circles that he was a skeptic is baseless. His skepticism, although it may have appeared significant to his orthodox theological opponents, was as peripheral as that of Aristotle or Locke. What kind of skeptic retreats to an assertion of "faith"?

George Wilhelm Hegel (1770–1831), perceiving the problems of logically sustaining this static idealism which, in effect, ruled out development, created a new, "dialectical" idealism. Like Plato he envisaged a God (the Absolute, the Idea, the Spirit) of which nature (the Other) is a reflection: God discloses himself "in a sheer other of Himself." This Absolute exists beyond time, in eternity, and again like Plato's One, is akin to thought: "The Spirit is declared in the element of pure thought." Although the Absolute is itself unchanging it manifests itself in the Other in the form of development: "the whole is the essence perfecting itself through its development." If we are to understand God in his eternal essence, we must break with the old formal logic of "right thoughts" and go back to the Greek dialectical method, seeing not simple cause and effect relations but "contradictions" endlessly manifesting themselves in the (Socratic) threefold process of thesis, antithesis (anti-thesis) and synthesis. In the process some ideas are "negated" but something of them remains in a resulting "higher" form, both the "higher" and the remnants of the lower existing in a "unity":

> Because the result, the negation is a specific negation, it has a content. It is a fresh notion, but higher and richer than its predecessor; for it is rich by the negation of the opposite of the latter, therefore contains it but also something more, and this is the unity of itself and its opposite. It is this way that the system of notions as such has to be formed—and has to complete itself in a purely continuous course.[43]

This "dialectical" logic is not simply a character of the mind but permeates the "soul" of nature: "this dialectic is not an activity of subjective thinking applied to some matter externally, but is rather the matter's very soul putting forward its branches and fruit organically." "Matter's very soul," is not, however, as it might appear at a first reading, something inherent in matter but the reflection of God in matter, matter being seen – as in Plato – as the antithesis of God (the Absolute). The primary emphasis is, as with all idealists, upon mind, divine or human (a reflection of the divine): "the Method is no-ways different from its object and content: for it is the content in itself, the *Dialectic,* which it has in itself, that moves it on." Within the context of this basically idealist approach – the Dialectic as God –, however, Hegel does attempt to undermine Kant's unchangeable and ever-unresolved categories, which, Hegel may have thought, left God as an unprovable hypothesis (in spite of "faith"). Hegel accepts categorical thinking but argues that the categories although contradictory blend and change. Truth can be grasped only by discarding static, compartmentalizing logic and substituting dialectical logic.[44]

Development in nature proceeds in accordance with the divine laws of the Dialectic, one of which – as Hegel follows Heraclitus – is contradiction: "contradiction is the root of all movement and life." Hegel lists a number of contradictions, interlocking opposites that, in spite of being opposites, have a basic identity in common:

> Positive and negative are supposed to express an absolute difference. The two however are at bottom the same: the name of either might be transformed to the other. Thus, for example, debts and assets are not two particular, self-subsisting species of property. What is negative to the debtor is positive to the creditor. A way to the East is also a way to the West. Positive and negative are therefore intrinsically conditioned by one another, and are only in relation to each other. The North Pole of the magnet cannot be without the South Pole and *vice-versa.* If we cut a magnet in two, we have not a North Pole in one piece and a South Pole in the other. Similarly, in electricity, the positive and the negative are not two diverse and independent fluids. In opposition the difference is not confronted by *any* other, but by *its* other.[45]

Something of this outlook had, as we have seen, been expressed by the Taoists, and like the Taoists, Hegel confusedly throws in very different phenomena, debtors and electricity, East and West. But he does illustrate the important point that things are to be studied not separately but in their interconnections, and he indicates the fluid nature of opposing phenomena. By implication he rejects a static view of nature or life and emphasizes development. Developmental blending, in fact, is central to the Hegelian outlook; for instance in regard to quantity and quality: "quality is implicitly quantity, and conversely quantity is implicitly quality. In the process of measure, therefore, these two pass into each other: each of them becomes what it already was implicitly." (Those acquainted only with Engels' presentations of

Hegel may be surprised at the degree to which he cleaned up Hegel's often obscure, even, at times mystical, views.) Hegel's formulations here obscure the basic point, namely that the new (quality) arises not from any supernatural source but from additions and combinations of the old.

Sometimes, Hegel noted, development takes place not gradually but suddenly:

> When people want to understand the rise or disappearance of anything, they usually imagine that they achieve comprehension through the medium of a conception of the gradual character of that rise or disappearance. However, changes in being take place, not only by a transition of one quantity into another, but also by a transition of qualitative differences into quantitative, and, on the contrary, by a transition that interrupts gradualness, and substitutes one phenomenon for another.

These sudden transformations take place, Hegel argued, at certain "nodal" points:

> Thus, for instance, the temperature of water is first of all indifferent in relation to its state as a liquid: but by increasing or decreasing the temperature of liquid water a point is reached at which this state or cohesion alters and the water becomes transformed on the one side into steam and on the other into ice.[46]

All these developments in philosophy took place in Europe. What was happening in Asia during these centuries? The answer is, nothing of consequence. Nor could it have been otherwise, for although there had been commercial development in Asia there had been no such rise of commercial capitalism as in Europe and no seizure of national power by commercial interests. Asia, still basically feudal, produced no Bacons or Diderots. Although Asian nations were, as the Japanese and Chinese port cities in particular showed, moving in the same directions as Europe, they moved at a slower pace. When this was later shaken up by imperialist aggression and new thinking began, its significant streams sometimes, as in China, took a socialistic direction.

The rise of commercial capitalism in Europe with a paid-labor base was a new historical phenomenon and produced new developments in philosophy: the inductive method and experimental orientation of Bacon, the science-influenced philosophy of Locke, the systematic skepticism of Hume, with its incisive, if negative, criticism of idealism, the militant materialism of Holbach and Diderot, the dialectical idealism of Hegel. Considerable though these advances were, however, they took place within the upper-class confines of a mass-exploitive society, and, hence, retained the overall limitations that had characterized feudal and slave-commercial thought. The fear of mass revolution was still ever present to consciously or unconsciously restrict and distort. Even the French materialists, although advocating social change, envisioned it as occurring within an exploitive (bourgeois) society, and failed to develop a materialist outlook for historical as well as natural development. Berkeley moved from the objective idealism of Plato to subjective idealism but it was still idealism and based upon the same premise – the primacy of Mind, cosmic or human. Clearly, if we view the universe as a delusionary surface whose wellspring is God, science or the advance of knowledge have little philosophical relevance. Hume developed a more sophisticated skepticism than had his Asian predecessors but again it was based on the sterile premise of the impossibility of understanding reality. The premises of idealism or skepticism once stated define forever the general boundaries of their systems of thought. Idealism and skepticism necessarily turn in circles with new abstractionist frills added periodically (Positivism, Existentialism). Materialism, on the other hand, can break out of the exploitive-society mold and continue to grow as knowledge grows.

All the philosophy we have surveyed so far was developed by male intellectuals and read and discussed almost entirely within the upper classes. The traditional dispute has been that between idealism – including religion – and the materialism and skepticism that arose from the commercial economic drive and opposition to landowning dominance. But below this ideological struggle between exploiting classes was a large if rather amorphous one between religion and mass skepticism (often with a materialist thrust). How extensive this current was is difficult to tell, for it was seldom put in written form and little research seems to have been done on it. But it is clear from the constant church inveighing in all ages and in all countries against "the faithless" that there must have been a great many of them. Even in present-day United States less than one-third of the population attends church and the indication is that in most civilized societies, the proportion has been and is much lower.

Nor should mass skepticism about religion cause surprise, for the doctrines of religion have always run counter to experience. The church proclaimed the existence of a God or gods but no God or gods were ever visible. It proclaimed divine providence but its proclamations were belied by the ever-recurring disasters of famine, flood, pestilence and war, which wiped out believers and non-believers alike. It proclaimed divine justice but the "good" suffered as much as the "evil." It proclaimed equality in the sight of God but there was no equality on earth; the masses toiled and suffered while their rich masters flourished. The church proclaimed the continuance of the soul after death but a dead person was clearly extinct. Even in hunter-gathering and early farming societies some must have had skeptical reservations similar to those later expressed in *Gilgamesh* and the *Rigvedas*. After the establishment of more developed societies, these reservations became part of the struggle of oppressed classes against their exploiters and their exploiters' church – coming to the surface in such massive events as the German and Chinese peasant wars or the French Revolution – a church which, until the rise of capitalism, was itself everywhere a greatlandowning organization employing peasant and slave labor.

The industrial working class, unlike the peasantry, was well-organized and literate, with its own newspapers, periodicals and pamphlets. Some Chartist and other 19th century workingclass leaders denounced "tithing" and other donations to the churches – "they preach to you to be content with your lot" – , and some took up the anti-clerical views of Voltaire and Thomas Paine. Some workers, such as Richard Carlile in Britain and Joseph Dietzgen in Germany, developed these views in their own way, Dietzgen formulating a materialist philosophy. How much response these views had among the workingclass, it is difficult to say, for then as now the class was riddled with religious propaganda, and people in exploitive-class societies hesitate to express skeptic or materialist views. It was with this class, however, that the next development of materialism lay.[47]

Chapter Two

MARX AND ENGELS

"Modern Materialism"

When the young Marx and Engels came together in Brussels in 1845 – three years before the *Communist Manifesto* – to formulate their world-view, they were concerned with philosophical as well as social matters. Marx had written his doctoral dissertation on the Greek materialists (in 1841) and both he and Engels had studied the works of Hegel and other contemporary philosophers. Early in the manuscript that they produced jointly, *The German Ideology,* they made a distinction between consciousness and ideology that Marx later elaborated. Consciousness reflects total existence: "Life is not determined by consciousness but consciousness by life." Within consciousness there is a special body of general ideas – on "politics, laws, morality, religion, metaphysics, etc." which in class societies reflect class interests: "The ideas of the ruling class are in every epoch the ruling ideas... The class which has the means of material [i.e., economic] production at its disposal has control at the same time over the means of mental production." Unlike previous philosophers they saw philosophical ideas not primarily as an individual but as a social phenomenon, as much, although less directly, an expression of class outlook as are social views.

As capitalism emerged from feudalism in Europe, certain general ideas – political, philosophical and so on – had, Engels noted, been formed in a kind of three-cornered class struggle.

> But side by side with the antagonism between the feudal nobility and the bourgeoisie was the general antagonism between the exploiters and the exploited, the rich idlers and the toiling poor.

In the early 19th century, with the growth of the industrial revolution the commercial-capitalist working class changed into an industrial-capitalism working class with the change from wind, wood and water power to coal, iron and steam. The social views of the new class, Engels noted, had been partly grasped by a small group of German and other skilled craftsmen in and around the Communist League and more deeply formulated by the industrial workers in the massive Chartist movement in Britain, where industry was much more developed than on the continent. Both groups perceived that capitalism was based on economic exploitation, and the Chartists advocated a social order run by the working class. "The working class," the Chartist leader, Bronterre O'Brien, noted in 1833 (when Marx was but fifteen), " . . . aspire to be at the top instead of the bottom of society – or rather that there should be no bottom or top at all."[1]

Marx and Engels, then, did not create Marxism out of whole cloth, but developed the already emerging views of the new class and formed them into a social science.

Although, as we saw, there were anti-clerical tendencies in the Chartist movement, stemming in particular from Thomas Paine, there was no materialist philosophy in early working class circles paralleling their advances in social thought. (Paine for all his ascerbic ridiculing of Christian theology, still believed in "one God" and "happiness beyond this life.") On the Continent the anti-religious views of the 18th-century materialists and skeptics penetrated some working class radical circles but again no consistent philosophical materialism emerged. In attempting to form such a philosophy, Marx and Engels had to rely mainly on the commerce-based materialists, from the ancient Greeks and Lucretius to Holbach and Diderot. These materialist views they combined with the dialectical way of thought, particularly as developed by Hegel. That their new materialism, however, no matter what its direct cultural roots, embodied the potentials for a working-class philosophy rooted in its condition of work, has been demonstrated in the present century, for wherever the working class has risen to power it has attempted to establish Marxist materialism along with Marxist social views.[2]

Although Marx and Engels realized the class roots of their world-view they did not believe that this affected its validity, either in social thought or philosophy. All such views were, in civilized societies, class conditioned. Yet truth had been discovered. The fact that Locke expressed the outlook of a rising commercial bourgeoisie did not mean that his argument on knowledge, arising not from a Platonic Godhead but from the senses, was untrue. Every class strove to discover and use whatever truths were useful to it. In this regard the industrial working class had one basic advantage. Being neither an exploiting nor a ruling class, it did not need the intricate web of lies necessary for exploitive rule and so could probe more deeply into some aspects of reality (as the Chartists had demonstrated in their social thought). But it also had disadvantages. The creation of ideology –

by historians, philosophers and others – was not one of its trades. And although it was literate, it was shut out from the wide-ranging education needed to form a truly sophisticated world-view. Such a view could be formed only by intellectuals allied with it, intellectuals who could rise above the incessant exploitive-class barrage of lies, prejudices and superstitions.

When Marx and Engels discussed their materialist philosophy they stressed what was new in it and paid scant attention to previous materialism, taking it for granted that their readers were acquainted with those arguments. When Marx declared (in 1868), "I am a materialist and Hegel is an idealist," he implied that he subscribed to a body of materialist views then generally known and, as a result of the publication of Darwin's *Origin of Species* in 1859, under active discussion: the universe consists only of matter existing in time and space; matter is eternal, and, hence, was never "created"; motion is an inherent part of matter; there is no Prime Mover, no God, and no "eternity" existing outside time and space; the universe arose from chance formations of atoms; it acts through objective laws that are discoverable by science; life emerged from matter and – including "Man" and his "mind" – evolved by natural selection; there is no "soul"; thought is a function of the brain; the mind, based on the brain, dies with the body; religion propagates delusion; there is no immortality, no Heaven or˙Hell; morality is a social phenomenon and does not depend on religious belief. The implication of these views, of course, is that religion is a delusion. As materialists, Marx and Engels were necessarily also atheists.[3]

Following their original collaboration in 1845, the development of the new materialism as science advanced fell almost entirely to Engels. This, as we shall see, was in some ways unfortunate, for Marx had a deeper grasp of fundamentals than Engels.

Engels began independently to expound their philosophy, emphasizing its relation to science. In response to an attack in 1877 by a German radical economist and philosopher, Eugen Karl Dühring (1833–1901). A revised version of part of this work, *Anti-Dühring,* was later published as *Socialism: Utopian and Scientific.*

In 1886 Engels wrote an exposition of the difference between "modern" materialism and the eclectic quasi-materialist, quasi-idealist views of Ludwig Feuerbach. Between 1872 and 1882 Engels put together a series of notes and essays on science and materialism that were published thirty years after his death under the title *Dialectics of Nature* (1925). Of these works, *Feuerbach* was written after the death of Marx (1883); and Marx apparently did not read the *Dialectics of*

Nature manuscripts; but *Anti-Dühring* was read to him by Engels and he wrote a chapter for it.[4]

Engels' overall view of the new materialism comes out succinctly in a passage in *Socialism, Utopian and Scientific:*

> Modern materialism embraces the more recent discoveries of natural science ... In both aspects [cosmic and biological evolution] modern materialism is essentially dialectic; and no longer requires the assistance of that sort of philosophy which, queen-like, pretended to rule the remaining mob of sciences. As soon as each special science is bound to make clear its position in the great totality of things and of our knowledge of things, a special science dealing with this totality is superfluous or unnecessary. That which still survives of all earlier philosophy is the science of thought and its laws—formal logic and dialectics. Everything else is subsumed in the positive science of nature and history.[5]

The philosophy with the deepest validity is "modern materialism." This materialism, based on science, is dialectical because nature, as the Greek materialists had first contended, is itself dialectical. Both cosmology and geology had now revealed evolutionary development, arising from conflictive interactions, the one in the universe, the other in living matter. Thinking based on these characteristics of nature is necessarily also dialectical, taking interactivity into account and using a fluid logic that encompasses more of reality than the traditional "formal" (either-or) logic. This more simplistic logic, however, still has its place in examining limited areas. Philosophy in the traditional sense will become obsolete (as the working class takes power) and there will no longer be a need for a special breed of thinkers called philosophers. A Marxist materialist can generalize from the findings of science, and this is really all there is to it. Idealist and skeptical philosophy will alike become obsolete, like alchemy or astrology.

Unlike idealism which, with its "eternal verities," has necessarily developed little since its inception, materialism has to "change its form" with "each epoch-making discovery" in science. By this Engels did not mean that the basic outlook has to change but only the concepts that are affected by the discoveries. In time, of course, these could be considerable and could affect some general perspectives. Materialism is developmental by its very nature. Holbach and Diderot did not simply follow Lucretius but incorporated the science of the day into their materialism. And Shelley, as we saw, included the newly discovered electrical "fluid" as matter.

In *Feuerbach,* Engels, exploring "modern materialism," singled out three major scientific advances: the law of the conservation and transformation of energy, the discovery of the biological cell, and Darwinian evolution to demonstrate how materialism was enriched by 19th century science. In discussing materialism today we have, of course, to advance beyond these discoveries into those of 20th century science. But, as the history of science shows, its earlier advances are not simply transcended but are incorporated into new advances. And by the 19th century certain basic facts had emerged which, although today placed in new perspectives, still have basic validity. Engels was, of course, not a natural scientist, but as a dialectical materialist making use of this new threshold of science, he had a wider grasp of the philosophical significance of scientific discoveries than non-Marxist scientists (as J.B.S. Haldane and other Marxist scientists have testified). Although major adjustments are needed, he still enables us to see deep into the nature of nature.[6]

The conservation and transformation of energy, Engels wrote, meant that "mechanical force, heat, radiation (light and radiant heat), electricity, magnetism" could be transformed one into the other with no loss of total energy because they are "modes of existence of one and the same energy, i.e., motion." Although Engels' formulation is dated, the "law" as such not only still holds good but it has been found to apply also on the atomic and sub-atomic levels. "Radiation," as Engels did *not* know, consists of elementary particles – photons – which in different wave lengths make up all the forms of "light," visible light as well as X-rays. When photons arise from the interactions of electrons and other elementary particles, the total energy in the particles involved remains the same before and after the transformation. Thus the "law" actually has deeper application than Engels realized. It is, however, not based on "motion" but on the nature of elementary particles and atoms. And the scope of the "law" has been widened. It is not restricted to energy but includes matter in general (mass plus energy). Engels, then, was wrong on specifics but right in his general contention that this process is universal and in doing so he gives us insight into the general nature of matter.

To Holbach and Diderot, as to Lucretius, the viable units of living matter were the same as those of matter, namely "atoms." But in the 1840's two German scientists, Theodor Schwann and Matthias Schleiden, discovered the biological cell, and Engels rejoiced:

> The hitherto incomprehensible miracle [of life] resolved itself into a process taking place according to a law essentially identical for all multi-cellular organisms.

Again, 20th century science, building on this foundation, has gone beyond it and has demonstrated the oneness of life on an even more elementary level than the cell, namely in the "chain" molecules that make up the cell, thus further supporting Engels' contention that the "miracle" can be unravelled in materialist terms.[7]

Darwin's theory of evolution by "natural selection" likewise is still accepted even as its genetic base – unknown to Darwin – is being revealed. Marx and Engels hailed the theory not only as providing an answer to the diversity of species but also to the basic philosophical puzzle of the nature and origin of the human mind.

Evolution and the Mind

On reading *The Origin of Species* (1859) Marx wrote: "Darwin's theory is very important and serves me as a basis in natural science for the class struggle in history." The theory, he felt, dealt a "death-blow" to Aristotelean "teleology" in the "natural sciences" (the view that all things are moving toward a predestined "goal"). In doing so, of course, it also dealt a blow to theology. And Engels noted that the theory supplied a vital link missing in previous materialism:

> The evolutionary series of organisms from few and simple to increasingly manifold and complex forms, as we see them today before our eyes, right up to and including man himself, has been proved in all its main basic features. Thereby not only has an explanation been made possible for the existing stock of the organic products of nature, but the basis has been given for the prehistory of the human mind, for following all its various stages of evolution from the protoplasm, simple and structureless yet responsive to stimuli, of the lower organisms right up to the thinking human brain. Without this pre-history, however, the existence of the thinking human brain remains a miracle.[8]

Neither Lucretius, Diderot nor Holbach really had any explanation for the human mind. True, Diderot in his "egg" analogy had glimpsed something of an evolutionary view of mind but it was a glimpse only. "Man" was generally presented as being eternal like the universe, and thought as a process similar to fermentation. Nor could it have been otherwise, for until a scientific theory of evolution emerged, no specific explanations were possible. Now, Engels implied, the idealist bugbear of a unique mind substance unrelated to matter (or the body) is finally laid to rest. If people evolved from simple biological organisms their brains must have evolved with them, from mere irritability in one-celled organisms to a certain reasoning quality in higher animals (including, Engels elsewhere noted, the fox), and, finally to conscious thought in humans. Evolution had given materialism a new major base.[9]

In his essay, *The Part Played by Labour in the Transition from Ape to Man* (1876), Engels argued that physiological change underlay mental evolution:

> Climbing assigns different functions to the hands and the feet, and when their mode of life involved locomotion on level ground, these apes gradually got out of the habit of using their hands [in walking—Tr.] and adopted a more erect posture. *This was the decisive step in the transition from ape to man.*

Scientists claimed in Engels' day and, in fact, until quite recently, that the "decisive step" was the development of the brain; fossil discoveries, however, have shown that Engels' conjecture (apparently first suggested by Ernst Haeckel) was correct. Uprightness came first, as a prerequisite. Uprightness, Engels went on to argue, would free the hands for further development and this would in turn develop the brain; a line of thinking perhaps begun by Anaxoragus, who, according to Aristotle, said that mind arose from "the possession of hands." Engels, of course, had no knowledge of genetic processes and little of the time spans involved in such developments, but his general materialist outlook steered him in the right direction. The contention that there was a direct "transition from ape to man" long doubted by scientists, has been supported by discoveries that apes and humans differ in only one percent of their genetic material and that hominids and apes are comparatively close in evolutionary sequence; not, as had been thought, some 20 million years apart. The human brain, Engels argued, evolved from the ape brain and is reflected in "the mental development of the human child."[10]

Among the mental heritage of humans, Marx early (1844) concluded, were instincts:

> Man is directly a *natural being.* As a natural being and as living natural being he is on the one hand endowed with natural *powers of life* — he is an active natural being. These forces exist in him as tendencies and abilities—*as instincts.*

In trying to reason these matters out in *The German Ideology* (1846) Marx and Engels spoke of early "man" as having a kind of "herd consciousness" and speculated that at a certain point human "instincts" might have become "conscious." In view of recent controversies on these questions, we should note that neither Marx nor Engels felt that the human possession of instinct stood in opposition to their social theories. Both were valid, but in different spheres. Instincts, however, Marx saw not as absolutely determinant behavioral patterns but as "tendencies" (which could either be developed or weakened by social forces). Today, as we shall see in discussing modern science, some of the specifics in this field have begun to emerge.[11]

Darwin, Marx and Engels believed, had demonstrated that human life had evolved from animal life. But what of the origin of life itself? On this question 19th century science had made possible only a generalized materialist view:

> With regard to the origin of life, therefore, up to the present, science is only able to say with certainty that it must have arisen as a result of chemical action.

Today the outline of the process, which involves mainly some "two dozen molecules of moderate complexity," has begun to take shape. Engels, following the science of the day, suggested that carbon compounds must have played a vital role, and this has been shown to be correct. Carbon atoms, with their capacity to "bind" other atoms, provide a key ingredient for combining protein molecules into long "chain molecules."[12]

Engels, as we can tell from his letters and *Dialectics of Nature,* did a great deal of reading in science. This raises an interesting question: Why did not the scientists themselves develop a rounded materialist philosophy? Darwin, for instance, although he privately confessed himself a materialist, was content simply to expound his theory; and his leading supporter, Thomas Henry Huxley, achieved no more than a watered-down Humean skepticism ("agnosticism," which Engels dubbed "shamed-faced materialism"). Most 19th century scientists, in fact, like Newton before them, accepted religious views of one kind or another. The problem is obviously not simply one of intelligence, but of formative social forces. Engels was probably no more inherently "intelligent" than Huxley. But Darwin and Huxley were members of the British upper classes and felt the restraining hand of these classes, which permitted them to do their scientific work but placed taboos upon generalizations that ran counter to the interests of the existing order and its religion. This restriction was sometimes direct, as in the open attacks on Darwin and Huxley by their religious opponents, but more commonly, and more pervasively, it was something simply "understood."[13]

As we have seen, materialism emerged over the centuries out of the capitalist struggle against landholding interests. With the feudal defeat and the capitalists' rise to ruling class in 19th century Europe, the tendency towards materialism emanating from the capitalist class virtually ceased, and the tendency received a new social base. The main task of developing materialism beyond Diderot and Holbach fell to Marx and Engels.

Truth and Practice

With the subjectivist trend in philosophy begun by Déscartes and continued by Locke, Berkeley and Hume, the problem of "knowing," epistemology, began to be a major concern of philosophers. If mind and matter are antithetical, how can mind know matter? Can the mind uncover "truth"? Attempts at a solution led to such desperate guesses as Leibnitz's that mind and matter, although unique entities, are like two separate clocks ticking in synchronization. Berkeley's solution that *all* – internal or external – is mind was reduced to solipsism ("I am the world") by Hume. Kant's structuralization of the mind into categories meant that all we can know are the categories. Reality, the "thing in itself," remained forever beyond us. The answer of materialism, implicit at least from Lucretius on – "mind and spirit are both composed of matter" – is that as mind is a form of matter, it can interact with matter.

To this general materialist view Marx and Engels added the evolutionary perspective, from "the lower organisms right up to the thinking human brain," which suggests that the brain has the capacity to reflect nature because it grew out of it and is still part of it.

Even before this evolutionary perspective was clear, Marx had attacked the basis of traditional epistemology in 1845:

> The chief defect of all hitherto existing materialism—that of Feuerbach included—is that things, reality, sensuousness, are conceived only in the form of the *object,* or of *contemplation,* but not as *human sensuous activity, practice,* not subjectively.

> The question whether objective truth can be attributed to human thinking is not a question of theory but is a practical question. Man must prove the truth, i.e., the reality and power, the this-worldliness of his thinking in practice. The dispute over the reality or non-reality of thinking which isolates itself from practice is a purely scholastic question.[14]

Brief and unadorned though these notes (on Feuerbach) are, they have basic philosophical implication. (Engels later titled the notes "Theses on Feuerbach," but Marx called them simply "ad Feuerbach" (on Feuerbach). "Theses" gives an air of dogmatism that Marx did not intend. In the first note, Marx renounces the abstractionist concept of mind as a contemplative machine. We "know" not just because we think but also because we act and feel. Although this dialectical blending has, as we shall see, been given scientific basis by modern brain, hormone and psychological research, bourgeois philosophers still attempt to solve the problem of "knowing" by presuming an abstract construct of mind as their theoretical base. What Marx is saying is that there is in actuality no "mind here" and "reality there" but a fluid interaction of related elements – physical matter, body and mind, both also forms of matter.

Most people use practice as a rough criterion of truth in their daily work and living. Marx is here proposing it as a general philosophical principle. We are able to grasp objective reality not only because we think and feel and act but also because we explore and test. The two notes, in short, lay the base for a scientific explanation for "knowing."

Engels later (1880) tackled the epistemological question in a sardonic attack on agnosticism (skepticism) in his introduction to *Socialism, Utopian and Scientific*:

Again, our agnostic admits that all our knowledge is based upon the information imparted to us by our senses. But, he adds, how do we know that our senses give us correct representations of the objects we perceive through them? And he proceeds to inform us that, whenever he speaks of objects or their qualities, he does in reality not mean these objects and qualities, of which he cannot know anything for certain, but merely the impressions which they have produced on his senses. Now, this line of reasoning seems undoubtedly hard to beat by mere argumentation. But before there was argumentation, there was action. *Im Anfang war die That.* And human action had solved the difficulty long before human ingenuity invented it. The proof of the pudding is in the eating. From the moment we turn to our own use these objects, according to the qualities we perceive in them, we put to an infallible test the correctness or otherwise of our sense perceptions. If these perceptions have been wrong, then our estimate of the use to which an object can be turned must also be wrong, and our attempt must fail. But if we succeed in accomplishing our aim, if we find that the object does agree with our idea of it, and does answer the purpose we intended it for, then that is positive proof that our perceptions of it and of its qualities, so far, agree with reality outside ourselves.[15]

As the "so far" indicates, Engels is not claiming that "practice" provides an absolute criterion for truth, only that it provides a basic one. But he did not explore the question further (as Lenin did). There are clearly levels of practice; and Engels here is speaking mainly of the general correspondence between objective reality and our concept of it. He was also, of course, aware that scientific practice reveals truths not visible to the senses unaided by the instruments of science, the kind of truths that Marx had in mind when he wrote: "Scientific truth is always paradox if judged by everyday experience, which catches only the delusive appearance of things."[16]

By "practice," Marx and Engels presumably mean both social struggle and scientific investigation, including experimentation. They do not make it clear, however, although they were certainly aware of it, that while practice is the ultimate criterion of truth, it is not always a direct criterion. When, for instance, they formed their economic and political theories they were not solely or even mainly depending on their own experience but on a written body of knowledge that had roots in the experience of others. So, too, with natural science. The views of a physicist are based not only on experiments but on general theories of physics which themselves may be based only ultimately on experiments or other forms of verification (such as astronomical observation). Even when theory, as with Einstein, precedes experimental verification, it could not have been conceived without a base in previous verifications.

Marx's final note on Feuerbach's philosophy, brief though it is, also casts a wide net:

> The philosophers have only interpreted the world in various ways: the point however is to *change* it.

The time had come – as Marx saw events build to 1848, when radical thinking, philosophical as well as political, abstract as well as concrete, must be directed towards the emancipation of humanity. So, too, as appears from other comments by Marx and Engels, with the arts. In short, with the coming of industrial capitalism, revolutionary thought and art had acquired a new character, one arising from purposive, historical action.[17]

Of course, this had to some degree been true previously. Diderot, for instance, realized that his philosophical and social views could assist in overthrowing the feudal order and were, at least in part, formed by that struggle. So, too, did Voltaire and Rousseau. But the concept was but fragmentarily inherent in a body of thought without a scientific social perspective.

One of the goals of Kantian and other philosophies had traditionally been that of discovering "absolute" truth in "an aggregate of finished dogmatic statements." On the contrary, Engels argues, knowledge must "remain relative for long successions of generations and can be completed only step by step" by science, social and natural. Such steps, the idealists contend, reveal only the surface of reality: "The number and succession of hypotheses supplanting one another . . . gives rise to the idea that we cannot know the *essence* of things (Haller and Goethe)." Albrecht von Haller had contended that we can only know the "outer shell" of nature. Goethe answered:

> *Nor shell nor kernel Nature does possess,*
> *Is everything at once and nothing less.*

Nature, that is to say, is a continuum whose various levels are all real and all interconnected. With this view Engels agrees. Yet, Engels continues, although science can penetrate to "the essence of things," there is really no "absolute truth": "Dialectical philosophy dissolves all conceptions of final, absolute truth." We should search instead for "relative truths along the path of the positive sciences." These truths, as they mount, point us towards "the essence of things."[18]

Although Marx and Engels discuss science as a means of uncovering reality, they fail to discuss the arts in this context. True, the arts, unlike science, are not centered on knowledge, but they do express views and make a unique contribution to knowledge. Imaginatively and emotionally as well as intellectually they project the essence of reality: in narrative, character, metaphor, color, movement, design, sound and so on. And this form of expression gives people insights into reality not supplied by the sciences or philosophy. In poets like Lucretius and Shelley, who were steeped in the science of their respective eras, this can be particularly illuminating. For instance, Shelley in *Hellas* visions the reactionary power of the Turkish empire about to attack Greece (in 1821) as perched "on the trembling pyramid of night." This refers to the fact that the earth casts a long, wavering, conelike shadow into sunlit space. But the vividness of the image both gives us an understanding of this beyond that of expository description and drives home the dark horror of the assault. The "bold, thundering song" of Lucretius's exposition of science and materialism impressed Marx. Even writers and other creative artists who do not have these broad interests nevertheless observe life and record depths not generally expressed.

When such insights are in line with scientific views, natural or social, they can not only greatly enhance them but can imprint their meaning in the human mind more deeply than can exposition alone. These insights, too, should be integrated into the Marxist world-view.

If we pause now and look back at the contributions by Marx and Engels to materialism, it is clear that they advanced it enormously beyond its previous high point, that of Diderot and Holbach. They gave it a greater scope of insight into reality, building on the new findings of science. Most significant of all they extended it to the analysis of society. "Old materialism," Engels wrote,

> looked upon all previous history as a crude heap of irrationality and violence; modern materialism sees in it the process of the evolution of humanity, and aims at discovering the laws thereof.

Moreover, as Engels noted, "modern materialism" was not just "materialist," it was also "dialectical."[19]

"Scientific Dialectics"

Although compared to Engels, Marx made few comments on dialectics, his are, like those on materialism and epistemology, of basic significance. Furthermore, some of them are less central to Engels' thinking and consequently have not always been properly integrated into later views of dialectical materialism, which have generally been based more on Engels than on Marx.

In his *Afterword* to the second German edition of *Capital* (1873) Marx quoted from a review of the first edition:

> "The one thing which is of moment to Marx, is to find the law of the phenomena with whose investigation he is concerned . . . Of still greater moment to him is the law of their variation, of their development, i.e., of their transition from one form into another, from one series of connexions into a different one."

Marx commented:

> Whilst the writer pictures what he takes to be actually my method, in this striking and (as far as concerns my own application of it) generous way, what else is he picturing but the dialectic method?
>
> Of course the method of presentation must differ in form from that of inquiry. The latter has to appropriate the material in detail, to

analyse its different forms of development, to trace out their inner connexion. Only after this work is done, can the actual movement be adequately described. If this is done successfully, if the life of the subject-matter is ideally reflected as in a mirror, then it may appear as if we had before us a mere *a priori* construction.[20]

The first thing a dialectical materialist thinker has to do, then, is to get the facts straight: "appropriate the material in detail." "Inquiry" – the analysis of these raw facts – comes next. Without this foundation, Marx implies, we have theory in an intellectual vacuum. The "clue" to historical "phenomena," he noted in a letter in 1877, must be found by specific research; and he added: "one will never arrive there by the universal passport of a general historico-philosophical theory, the supreme virtue of which consists in being super-historical [above history]." Unless facts are first ascertained by research or struggle one can fall, Marx sardonically notes, into pretentious idealist dialectics:

> Wherein does the movement of pure reason consist? In posing itself, opposing itself, composing itself; in formulating itself as thesis, antithesis, synthesis; or, yet again, in affirming itself, negating itself and negating its negation.

Through research, such as Marx expended on *Capital,* we first find the "details" of the phenomena being investigated and then analyze them and discover their "forms of development" and "inner connections." Only then can the total "movement" be "presented." It might seem, he notes, that this picture represents general forms of thought being imposed on reality – "a mere *a priori* construction" – but this is not so. The "inner connections" have objective existence because society acts in dialectical ways. These include "development" and "transitions from one form into another" (as Marx quotes the reviewer with approval). And, as he makes clear in a number of places, they spring from "contradiction." Society (like nature) is fluidly conflictive in its essence: "What constitutes dialectical movement is the existence of two contradictory sides, their conflict and their fusion into a new category."[21]

Although something of the technique of finding interconnections had been uncovered by Hegel, Hegel's was a technique in a philosophical fantasyland:

> My dialectical method, is not only different from the Hegelian, but is its direct opposite. To Hegel, the life-process of the human brain, i.e., the process of thinking, which, under the name of 'the Idea,' he even transforms into an independent subject, is the demiurgos [creator] of

the real world, and the real world is only the external, phenomenal form of 'the Idea.' With me, on the contrary, the ideal is nothing else than the material world reflected by the human mind and translated into forms of thought.[22]

Marx's meaning here becomes clearer in the light of a passage in Engels' *Feuerbach:* the Hegelian God (the "Absolute," the "Idea") " 'alienates' itself by changing into nature," then – in a negation of the negation – "comes again to self-consciousness in man." This is a mystical dialectical spinoff from Plato: the cosmic Mind, the antithesis of matter, suffuses matter; the human mind is a reflection of the cosmic Mind (the One). However, while the Platonic One is static and tends to produce stability in society and nature, the Hegelian Absolute produces change, using the human mind as its imperfect tool. Marx, then, is not speaking just of the human mind but also of the divine "Idea" (God) which idealists perceive as extending into the human mind. The "real world" is not, as Hegel contends, a reflection of this Idea; on the contrary, the Idea is a distorted reflection in the human mind of the "real" world, the "material" world. In the end, however, for all his mystical elaborations, Hegel's view, like Berkeley's, boils down to solipsism). "Thus, for Hegel, all that has happened and is still happening is only just what is happening in his own mind."[23]

By "dialectical method" here Marx clearly means Hegel's general approach to philosophical problems and not just his logic or method of reasoning. Hegel began with Mind and (delusively) saw a spiritual force behind the working of matter, the human mind and society. Marx began and ended with matter, society and people (a blending of the first two).

As a young student, Marx had been impressed by some aspects of Hegel, but he soon came to see that Hegel's political views were reactionary and later he attacked his view of history as the "old Hegelian junk" and denounced the "mystic semi-obscurity" of his philosophy. However, as he tells us in a letter of 1858, when he was preparing to work on *Capital,* "by mere accident I again glanced through Hegel's Logic" and found it of "great service . . . as regards the method of dealing with the material." And he added "I should like very much to make accessible to the ordinary human intelligence – in two or three printer's sheets – what is rational in the method which Hegel discovered but at the same time enveloped in mysticism." There is, then a "rational" core to Hegel's "logic", his "method of dealing with the material" (even though his general philosophical method was sheer "mysticism"). As we consider *Capital,* it becomes apparent that one "great service" Hegel's "method of dealing with the material" provided for Marx was that "tracing" of "interconnections" he noted in his comments on the review of *Capital*. These interconnections pointed towards the "movement" both of the capitalist system as a whole and its various parts. Marx also tells us that in *Capital* "in the chapter on the theory of value" he "coquetted with the modes of expression peculiar to him [Hegel]." Such expressions, presumably as that expressing "Hegel's discovery that merely quantitative turn into qualitative changes." Engels perceived this use of Hegel in *Capital* and wrote to Marx that he should have noted it more explicitly. Although Marx was clearly aware of these dialectical patterns his comments show that, in accordance with his "rational" method, he did not impose them on his material but used them as they emerged from it. He implies also that in his works to that time, from 1845 to 1858, he had not consciously used the Hegelian method of logic. He found it "of service" when he came to the broad canvas of *Capital* and there only in limited ways. It had, then, played no particular part in *The Communist Manifesto,* the *Notes on Feuerbach* or his political writings on the French (1848) and other situations. In *The Poverty of Philosophy* (1847), he ridiculed Hegel's philosophy at length. He certainly did not want to be identified with it; and this may be implied in the, rather derogatory, "coquetted."[24]

Hegel's idealist dialectics must be repudiated; and materialist, inherently revolutionary, dialectics asserted:

> In its mystified [Hegelian] form, dialectic became the fashion in Germany, because it seemed to transfigure and to glorify the existing state of things. In its rational form it is a scandal and abomination to bourgeoisdom and its doctrinaire professors, because it includes in its comprehension and affirmative recognition of the existing state of things, at the same time also, the recognition of the negation of that state, of its inevitable breaking up; because it regards every historically developed social form as in fluid movement, and therefore takes into account its transient nature not less than its momentary existence; because it lets nothing impose upon it, and is in its essence critical and revolutionary.
>
> The contradictions inherent in the movement of capitalist society impress themselves upon the practical bourgeois most strikingly in the changes of the periodic cycle, through which modern industry runs, and whose crowning point is the universal crisis.[25]

The bourgeoisie has no objection to idealist dialectics. They welcome it because it can be used to contrast unrealizable ideal development with an implied unchangeable social order. Hegel, in fact, supported the feudal Prussian state in his *Philosophy of Right* (1821). It is only when dialectics becomes "rational" and is applied to historical phenomena that the attack begins. The "existing state of things," is not, as the professors of "bourgeoisdom" argue, eternal, but transient; and this includes the capitalist system itself. Social systems, the "rational" dialecticians point out, collapse because they are torn apart by internal "contradictions." So, too, capitalism, with its boom-and-crash cycles leading to a "universal crisis."

What Marx meant by the "rational" or "scientific" dialectic is further clarified in his comments on the French bourgeois intellectual, Pierre Joseph Proudhon (1809-1865):

> For an estimate of his book [*The Philosophy of Poverty*], which is in two fat tomes, I must refer you to the work I wrote as a reply. There I showed, among other things, how little he had penetrated into the secret of scientific dialectics; how, on the other hand, he shares the illusions of speculative philosophy, *for instead of conceiving the economic categories as theoretical expressions of historical relations of production, corresponding to a particular stage of development of material production,* he garbles them into pre-existing, eternal ideas; and how in this roundabout way he arrives once more at the standpoint of bourgeois economy.

This is based on Hegel:

> He [Hegel] thinks he is constructing the world by the movement of thought, whereas he is merely reconstructing systematically and classifying by the absolute method the thoughts which are in the minds of all.

The idealist dialectical thinker, then, although he may perceive movement and interconnections perceives them as embodying abstractions which shape reality. Thus Proudhon does not derive his "economic categories" from an examination of actual economic conditions; on the contrary he sees economic conditions as arising from "pre-existing, eternal" categories (much as bourgeois historians today see "freedom," "courage," "ambition," "greed," and so on as the controlling forces of history). As Marx sarcastically commented in *The Poverty of Philosophy*: "What Hegel has done for religion, law, etc., M. Proudhon seeks to do for political economy."

Marx continues:

> Proudhon had a natural inclination for dialectics. But as he never grasped really scientific dialectics he never got further than sophistry.

In fact this hung together with his petty-bourgeois point of view. Like the historian Raumer, the petty bourgeois is composed of on-the-one-hand and on-the-other-hand. This is so in his economic interests and therefore in his politics, in his religious, scientific and artistic views. So in his morals, in everything. He is a living contradiction. If, like Proudhon, he is in addition a clever man, he will soon learn to play with his own contradictions and develop them according to circumstances into striking, spectacular, now scandalous, now brilliant paradoxes.

A "clever" idealist thinker, then, can use what we might call "dialectics in a vacuum" to produce juggler-like illusions. He plays non-existent contradictions one against the other in a pyrotechnic display of "on-the-one-hand and on-the-other-hand" logic; which ends not only in confusion but reaction: "Charlatanism in science and accommodation in politics are inseparable from such a point of view."[26]

For Marx "scientific dialectics" was rooted in "practice" and a systematic analysis that uncovers total movements. At the essence of reality, natural and social, is "conflict" leading to "fusion." Dialectical thinking "is in its essence critical and revolutionary." "Scientific dialectics" provides a weapon whereby one "changes" society and challenges reactionary views in all fields, social and philosophical.

If we put Marx's various comments together, they form a cohesive whole. And dialectics is but one aspect of this whole. Marx did not believe that a mastery of dialectical thinking would provide a magical path to truth. Practice and not dialectics is at the root of his "method." It is revolutionary philosophy – "change" the world – that gives life to dialectical thinking. Without it, such thinking, as in Hegel or Proudhon, tends to reaction. The dialectical materialist is not primarily dialectical but materialist, and practice-based; not a closeted scholar but a revolutionary (who, like Marx himself, is perhaps also a scholar). Dialectical logic has not become the whole of or even the main base for thinking even though it reveals interconnections and movements that cannot otherwise be fully grasped. More important, the implication runs, is inductive thinking on a materialist basis, the thinking, emphasized by Bacon, of reasoning from the particular to the general. Inductive reasoning is directly or ultimately based on practice which reflects the interaction of thought, sensation and action. When, however, we establish generalizations by this method they can be used (deductively) in the further search for facts and interconnections, as Marx demonstrated extensively in *Capital*. These interconnections are necessarily conflictive, arising from the "fusion" of the "contradictory sides" of phenomena and taking the form of quantity to quality change.

Engels: Insight and Error

Neither Marx nor Engels, as I have noted, used the term "dialectical materialism." In 1891 the Russian Marxist, Georgi Plekhanov spoke of "modern, dialectical materialism," apparently compressing Engels' 1877 statement in *Anti-Dühring:* "modern materialism is essentially dialectical." Then Plekhanov wrote in 1892: "The philosophy of Marx and Engels is not only a *materialistic philosophy:* it is *dialectical materialism.*" Engels contrasted "modern materialism" to the "mechanical" materialism of Diderot and Holbach, in whose age nothing was known of the interactivities of matter disclosed by the transformation of one form of "energy" into another; of Darwinian evolution or of the cellular units of life with their fluidity of interaction. It was "mechanical" (although, as we have seen, by no means static) because it was based on the Newtonian machinelike universe and only partly broke from the set patterns of animal species proclaimed by the Linnaean "naturalists." "Modern materialism" was also a materialism that, unlike previous materialisms, examined society as well as nature in materialist terms; and saw that it too, was evolutionary. As Engels not only spoke of "modern materialism" as being dialectical but used the term "dialectical philosophy"; "dialectical philosophy itself is nothing more than the mere reflection of this process ["of becoming and passing away"] in the thinking brain," he would probably have approved of "dialectical materialism." Lenin, following Plekhanov, used "dialectical materialism" in *Materialism and Empirio-Criticism.* In his *Granat Encyclopedia* essay on Marx, he also used the term, but he there divides Marx's philosophy into two sections, first, "Philosophical Materialism" and, second, "Dialectics." This was also the emphasis that Plekhanov intended: the "philosophy of Marx and Engels" is primarily "materialistic." It seems unlikely, however, that Marx would have accepted the term "dialectical materialism" for "dialectical" then implied dependence on Hegelian idealism, which he regarded as the reactionary antithesis of his own "scientific" and "revolutionary" dialectics.[27]

Engels, in restating the three "general laws" of dialectics emphasizes that they are not something imposed on nature or society but are inherent in the processes of both:

It is, therefore, from the history of nature and human society that the laws of dialectics are abstracted. For they are nothing but the most general laws of these two aspects of historical development, as well as of thought itself. And indeed they can be reduced in the main to three:

The law of the transformation of quantity into quality and vice versa; The law of the interpenetration of opposites; The law of the negation of the negation.

In *Dialectics of Nature* in particular, Engels discusses the working of these "laws" as illustrated by the science of his time. In preparation for this, as J.B.S. Haldane notes, he read widely in scientific works; and he was "in close touch" with a distinguished scientist, Karl Schorlemmer, of Owens College, Manchester, a good friend and fellow socialist, one of the founders of organic chemistry. It is, Haldane, concludes (in 1939), "astonishing how Engels anticipated the progress of science in the sixty years since he wrote."[28]

I shall consider first the "law" (the general question of "law" I shall discuss later) that Engels places second but which Marx considered basic, namely the "interpenetration of opposites":

Dialectics [Engels wrote], so-called *objective-dialectics,* prevails throughout nature, and so-called subjective dialectics, dialectical thought, is only the reflex of the movement in opposites which asserts itself everywhere in nature, and which by the continual conflict of the opposites and their final merging into one another, or into higher forms, determines the life of nature.[29]

"Everywhere in nature," then, there is a "conflict of the opposites" – or "contradiction." These "opposite" elements can simply merge "into one another," in which case the movement is simply circular; or they can produce "higher forms." They do not, then, inevitably produce such forms; that is to say, result in "development." The "laws" of dialectics that govern these phenomena are both objective and subjective; the objective ones are part of the processes of nature and society, the subjective ones are a reflection of these processes in the mind.

If these processes are continuous in nature and society, why are people generally not aware of them? Engels did not, directly pose the problem, but the answer he implies is that exploitive-class society engenders a "static" manner of thinking that postulates set entities rather than interpenetrative conflict as the essence of reality:

So long as we consider things as static and lifeless, each one by itself, alongside and after each other, it is true that we do not run up against any contradictions in them ... But the position is quite different as soon as we consider things in their motion, their change, their life, their reciprocal influence on one another. Then we immediately become involved in contradictions.[30]

In Engels' day neither the proton nor the electron, with their negative and positive charges, had been discovered, nor had the conflict between gravitational and electro-magnetic "reactions" (as in star evolution, with gravity pushing in and electro-magnetism pushing out). Still Engels argued against the then current concept that "attraction is a necessary property of matter, but not repulsion." "The essence of matter," he wrote, "is attraction and repulsion." Modern science has borne him out, both on the elementary particle level and in cosmic phenomena (derived ultimately from the nature of the particles).[31]

Development in living matter, like that in matter, is determined by the interaction of opposite forces:

> Life consists just precisely in this—that a living thing is at each moment itself and yet something else. Life is therefore also a contradiction which is present in things and processes themselves, and which constantly asserts and solves itself; and as soon as the contradiction ceases, life too comes to an end, and death steps in.[32]

Not only do matter and living matter act in response to contradictory processes, society does also: "the material productive forces of society come into conflict with the existing relations of production." This interpenetration of opposites produces class struggles, particularly, in capitalist society, that of the working class and capitalist class. And so, Engels noted, down into the details of social conflict: "It seems that every worker's party, in a great country, can only develop itself by internal struggle, and this is based on the laws of dialectical development in general."[33]

The conflict of opposites sometimes results in "equilibrium," not, however, a static but a changing equilibrium, as in "the motion of the heavenly bodies," or in living matter:

> Finally, in the living organism we see continual motion of all the smallest particles as well as of the larger organs, resulting in the continual equilibrium of the total organism during the normal period of life, which yet always remains in motion, the living unity of motion and equilibrium. All equilibrium is only *relative* and *temporary*.

The general picture here has been adopted by Stephen Jay Gould and other scientists today in their "punctuated equilibria" theories. We find the same phenomenon also in society, when antagonistic classes exist in a state of active equilibrium prior to a revolutionary surge (as in Czarist Russia or Batista's Cuba).[34]

Although Engels' comments give us basic insights into the workings of nature, some of them lack a consistent materialist perspective. For instance, in the statement that "life" is a "contradiction" which is "present" in "things and processes." On the contrary the "things and processes," particularly the proteins and nucleic acids (then unknown) and the cell itself as a whole determine the nature of life. Similarly "internal struggle" in political parties or unions is not "based on the laws of dialectical development" but on conflicting social forces. Clearly there are two interlocking elements at work. General phenomena are built up out of specific phenomena and general characteristics penetrate specifics. Both must be taken into account. The question is which is basic and which derivative; and the materialist answer is clearly that the specific phenomena are basic. To see the general as basic is to veer toward idealism, as with Aristotle.

One of the most important concepts of Hegel, Engels believed, was the "great basic thought that the world is not to be comprehended as a complex of ready-made *things,* but as a complex of processes." Although Engels recognized the importance of the examination of "things," especially when science was in an early stage – "one had first to know what a particular thing was before one could observe the changes it was undergoing" – he sometimes places an exclusive emphasis on process: the world is "a complex of processes." Engels is, of course, right in saying that "things" (the word is used in a broad sense) should not be considered as "ready made." They have come about by change and are always changing; but they exist and they form the basis for process. There can be no dance without dancers.[35]

It is, of course, also true that it is not always easy to distinguish thing from process. An atom is at once thing and process. Nevertheless the distinction exists. Elementary particles, in or out of atoms, are things; their interactions are processes, almost inextricably blended though the two may be. Molecules, as the electron microscope clearly reveals, are things, things that consist both of other things – atoms, other molecules – and internal processes, and have size and shape. As usual with two interacting elements, both are necessary but only one is basic. To see process only as primary is, a gain to veer towards idealism.

The "process" passage in *Feuerbach* presents further problems, especially if it is considered alone, as it often is. The "complex of processes" of "the world," Engels continues, "go through an uninterrupted change of coming into being and passing away." Why? The answer, of course, is the interpenetration of opposites. But Engels does not say so and gives the impression of a simple rise and fall of phenomena as the essence of the dialectical process. This is suggested also in another passage: "the uninterrupted process of becoming and passing away, of endless ascendency from the lower to the higher." And in *Dialectics of Nature:* "the whole of nature...has its existence in eternal coming to being and passing away." This *Dialectics of Nature* passage was later quoted by Joseph Stalin as the basis for his own view in his essay, *Dialectical and Historical Materialism:*

> Contrary to metaphysics, dialectics holds that nature is not in a state of rest and immobility...but a state of constant movement and change, of continuous renewal and development, where something is always arising and developing, and something always disintegrating and dying away.

Stalin's failure to treat contradiction properly and his almost exclusive emphasis in his essay on the rise of the "new" and the decline of the "old" takes the heart out of dialectical materialism, and although Stalin did not intend this, opens a path to idealism. For what causes this rise and fall, this "renewal and development?" God? Stalin's view has permeated many expositions of dialectical materialism, especially by Soviet philosophers, even as they were vilifying him.

For this confusion Engels, much though he would have deplored it, bears some responsibility. True, in other passages he emphasises contradiction but he does not do so consistently. It is not, for instance, central in his discussions of "motion." Furthermore we cannot conceive of Marx writing the *Feuerbach* pure "process" passages or suggesting that the essence of nature lay in "eternal coming into being and passing away." The "scientific" dialectic was "revolutionary in its essence." Engels certainly understood "the continual conflict of the opposites" intellectually but it did not, as with Marx, permeate his being. We might note, too, Engels' placing of the "law of the interpenetration of opposites" second in his listing, behind that of "the transformation of quantity into quality and vice versa." If we place this "law" first, the implication is left open that the change from quantity to quality causes "the interpenetration of opposites" and not the other way round. Again, we cannot conceive of Marx formulating things thus.[36]

There is, Engels points out, as he discusses the "law" that he puts in first place, no "miracle" involved in the creation of the new. All the variety of nature has come about by combinations of already existing entities: "qualitative changes can only occur by the quantitative addition or substraction of matter or motion (so-called energy)."[37]

The universal nature of the quantity-quality process is reflected in the periodic table of the elements:

> As in the case of oxygen: if three atoms unite into a molecule, instead of the usual two, we get ozone, a body which is very considerably different from ordinary oxygen in its odour and reactions. . . . How different laughing gas (nitrous oxide, N_2O) is from nitric anhydride (nitrogen pentoxide N_2O_5)! The first is a gas, the second at ordinary temperatures a solid crystalline substance. And yet the whole difference in composition is that the second contains five times as much oxygen as the first.

And Engels could have shown similar quantity-quality changes through the whole table, which lists the basic substances, solid, liquid or gas that combine to form natural phenomena. All these "elements" were clearly different from each other: hydrogen from iron, carbon from mercury, yet all owed these qualitative differences, that is to say, their unique identities, simply to quantitative differences: "atomic weights." (Engels would have been happy to learn that these "atomic weight" differences rested on the number of elementary particles within the atom.)[38]

As Engels contemplated the obvious nature of the phenomenon of quantity-quality change and its disregard by the scientists of his day, he commented, with wry humor:

> Probably the same gentlemen who up to now have decried the transformation of quantity into quality as mysticism and incomprehensible transcendentalism will now declare that it is indeed something quite self-evident, trivial, and common-place, which they have long employed, and so they have been taught nothing new. But to have formulated for the first time in its universally valid form (i.e., by Hegel) a general law of development of nature, society, and thought, will always remain an act of historic importance. And if these gentlemen have for years caused quantity and quality to be transformed into one another, without knowing what they did, then they will have to console themselves with Moliere's Monsieur Jourdain who had spoken prose all his life without having the slightest inkling of it.[39]

However, do all qualitative changes arise from the "addition or substraction of matter or motion?" Engels also points to another factor that is sometimes involved: "by means of a change of position and of connection with neighbouring molecules it ["the molecule"] can change the body into an allotrope or a different state of aggregation." An "allotropic" change is a change not between elements but within elements, arising from particular atomic or molecular formations. Graphite, charcoal and diamond are allotropic forms of carbon. Engels, then, is arguing that qualitative change can come about by means of "change of position" or, as he put it in another passage, "various groupings of the molecules" (a view anticipated in a general way by Lucretius). Engels apparently also believed that such factors were involved generally and not just in "allotropic" changes:

> All qualitative differences in nature rest on differences of chemical composition or on different quantities or forms of motion.

By "differences in chemical composition" Engels must mean the same thing as difference in molecular groupings and "change in position." (What he means by "forms" of motion, I will discuss later.) In a related passage, however, he writes as though quantity was the sole element involved: "qualitative changes can only come about by the quantitative addition or substraction of matter or motion." Perhaps, then, he regarded change in "position" as essentially quantitative, but he seems not to have grasped its full significance, which was not so clear then as now (for instance in the "double helix" of the nucleic acid molecular chains). Sometimes in qualitative change, only quantitative factors are involved, sometimes only positional ones, and sometimes both. In stating the general "law" involved, then, should we refer to it simply as that of quantity to quality or that of quantity-arrangement to quality? We can best examine this question after we have noted some of the discoveries of modern science.[40]

In some unpublished notes to *Anti-Dühring,* Engels (following Hegel) noted that "quality can become transformed into quantity just as much as quantity into quality . . . reciprocal action takes place." This is true enough, but to list it as though the two were of equal significance obscures – as in Hegel – the basic materialist point, namely that the new (qualitative change) arises by natural causes. Addition produces not just more of the same but something different. We need not invoke God or miracles or spirit, to explain the "creation" of the new.

"In spite of all intermediate steps, the transition from one form of motion to another always remains a leap, a decisive change." In such "decisive changes" there are "nodal points" between different forms of existence (as Engels again follows Hegel):

> For example, in the case of water which is heated or cooled, where boiling-point and freezing-point are the nodes at which—under nor-

mal pressure—the transition to a new form of aggregate takes place, and where consequently quantity becomes transformed into quality.

There are "nodal points" to mark the transition between the different forms of matter and between matter and living matter:

In the same way, the transition from the physics of molecules to the physics of atoms—chemistry—in turn involves a definite leap; and this is even more clearly the same in the transition from ordinary chemical action to the chemistry of albumen [protein] which we call life. Then within the sphere of life the leaps become ever more infrequent and imperceptible.[41]

Although Engels, following the science of his day, is wrong on specifics, his general argument is correct and has materialist implication. But we are not, as he suggests, dealing primarily with "forms" of "motion," and the "transition," is not from the "physics of molecules to the physics of atoms" but the reverse. Nevertheless there is a transition. Life is not only "the chemistry of albumen" (protein) but of the nucleic acids. Nevertheless it is essentially a chemical process. There are more forms of matter than Engels knew – from radiation to plasma. Nevertheless there are areas of overlap, "transition," between them; and it is this that creates overlaps between different specialties in physics and between physics and biology.

In "leap" Engels seems to imply not only a major change but also an increase in tempo; and this concept was later developed by Plekhanov. Modern science bears out the observation; things tend to speed up as a major change approaches. A star which has evolved for millions of years can collapse in a few minutes. Apes long working on a problem suddenly see its solution. Social change comes in revolutionary sweeps. Why the (diverse) parallels I shall discuss later.[42]

In *Anti-Dühring* Engels defended the validity of the third "law" of dialectics, the "negation of the negation," in considerable detail, arguing, as had Hegel before him, that all development went through a process in which one original entity was "negated" by another but continued within it and in the next stage emerged in a "higher" form. It was this ascent to a "higher" form – "higher and richer than its predecessor" – that Hegel, as an idealist, saw as the mystical center of the dialectical process: "the whole is the essence perfecting itself through its development." "A very simple process," Engels wrote, "which is taking place everywhere and every day, which any child can understand, as soon as it is stripped of the veil of mystery in which it was wrapped by the old idealist philosophy." Like Hegel, he saw manifestations of the "law" in nature, for instance, in plant reproduction:

> Let us take a grain of barley. Millions of such grains of barley are milled, boiled and brewed and then consumed. But if such a grain of barley meets with conditions which for it are normal, if it falls on suitable soil, then under the influence of heat and moisture a specific change takes place, it germinates; the grain as such ceases to exist, it is negated, and in its place appears the plant which has arisen from it, the negation of the grain. But what is the normal life-process of this plant? It grows, flowers, is fertilised and finally once more produces grains of barley, and as soon as these have ripened the stalk dies, is in its turn negated. As a result of this negation of the negation we have once again the original grain of barley, but not as a simple unit, but ten, twenty or thirty fold. Species of grain change extremely slowly, and so the barley of today is almost the same as it was a century ago.

Engels continues:

> But if we take an ornamental plant which can be modified in cultivation, for example, a dahlia or an orchid: if we treat the seed and the plant which grows from it as a gardener does, we get as the result of this negation of the negation not only more seeds, but also qualitatively better seeds, which produce more beautiful flowers, and each fresh repetition of this process, each repeated negation of the negation increases this improvement.[43]

These passages are, it seems to me, far from clear. Certainly they are not something "any child can understand." Engels seems to argue in the first part of the first passage that the sequence seed-plant-seeds is itself a negation of the negation. But, if so, where is the higher level of development? There appears to be none, only a change from one quantity to another, from one seed of barley to many. Then in the final sentence of this barley passage he seems to argue that this higher level is something that will arise from a large number of these quantitative changes in a "century" or more. But why should this be so? Accumulated repetition will not produce evolutionary change.

That it is, indeed, "higher level" development that Engels sees as the heart of the "negation of the negation" is clear from the orchid passage. But again the so-called "higher level" does not involve a negation of the negation, as Engels argues. If it exists at all it is a simple quantity-quality change arising from a deliberate mingling of gene pools (and not by mutation). But who is to say that this mingling, in fact, produces a "higher level," that the new flowers are indeed "more beautiful" than the old? They are simply different. Engels, in short, does not demonstrate that the negation of the negation is a developmental phenomenon, a "law" bringing about a "higher" form of the original entity. In fact, the more he talks about it, the clearer it becomes that it is essentially an idealist concept – as it is in Hegel. It should simply be dropped.

The laws of dialectics, Engels wrote in *Dialectics of Nature,* "can be reduced in the main to three." The implication seems to be that in addition to the three "main" laws there may be secondary ones. Could one of these be the law of the transformation and conservation of energy (actually, of matter) which Engels hailed as a major scientific discovery? He perhaps had this "law" in mind when he wrote that "qualitative differences in nature rest" not only on "differences" in quantity or "chemical composition" but on differences in "forms of motion" (from heat to electricity for example). This law, he clearly did not believe should be included under the "main laws" of dialectics:

> In the present work dialectics is conceived as the science of the most general laws of all motion. Therein is included that their laws must be equally valid for motion in nature and human history and for the motion of thought.

This particular law, Engels indicates, concerns only matter and living matter, not society or thought. It is "one of the general laws of nature," similar to the law of gravity. Such laws embody "different phenomenal forms of the same universal motion." However, as the objective laws of dialectics are also laws of nature, the implication of Engels' statements appears to be that those laws of nature that have wide application but do not apply to society or thought can also be regarded as secondary laws of dialectics.

All processes become laws when they are of a general character: "The form of universality in nature is *law.*" It is apparent from Engels' comments that "universality" can refer to general spheres of reality as well as to the whole of it. The law of gravity has universality in regard to "clumped" bodies of matter. But not to elementary particles or to social development.

So far, for the purpose of exposition, I have accepted this concept of "objective" law. The word law, however, should not, it seems to me, be used in this sense but should be restricted to human concepts ("subjective" law). If we regard the universe as materialists it is apparent that there are no laws inherent in nature. There are only things and processes (including different "reactions" – nuclear, electromagnetic etc.). There is no "law" of gravity; there is only gravity (with its own "reaction," connected with the interactions of particles whose nature can be deduced but which have not yet been detected). Similarly there are no objective "laws" of dialectics – in either nature, society or thought. What are called laws are in reality general processes arising from the interpenetration of opposite elements, natural, social and psychological. The "laws" are formulations of our understanding of these processes.[44]

So, too, with Engels' contention that "universality" of process is "law." It is simply universality of process, either in nature or society. We have natural process and social process. Some of it results in development but most of it does not. True, one set of processes can make others inevitable – barring outside interference – as in embryonic development or the evolution of stars. But no developments in nature or society were ever determined by "law." They arose out of the interactions of natural or social opposite elements: electro-magnetism and gravity, the class struggle.

Not only, however, is it clear that there is no objective law, the concept itself has idealist implications. A universe run by Law is much the same as one run by God. Engels, of course, did not intend this. He is simply using the Newtonian language still current in 19th century philosophy and science, without perceiving its implications. But the concept, like that of the negation of the negation, should be dropped from Marxist thought.

Engels argues that quantitative change comes about not only by the addition or substraction of "matter" but also of "motion (so-called energy)." Motion – "the mode of existence of matter" – he considered a more general phenomenon than energy:

> The expression "energy" by no means correctly expresses all the relationships of motion, for it comprehends only one aspect, the action but not the reaction. It still makes it appear as if "energy" was something external to matter, something implanted in it.

Engels was using "energy" in a sense then current: "energy is the term used for repulsion." "Force" was used for "attraction." "Motion" he used to include both.[45]

Clearly this nomenclature has gone by the boards along with the concepts behind it. Energy is now seen as one of the two basic components of matter (the other being mass). And matter includes both repulsion and attraction; for instance, negative and positive electrical charges.

Engels regarded motion as "the mode of existence of matter."

"Heat, light, electricity, magnetism" are "forms of motion." The difference between them lies in a "transition from one form of motion to another." All this is clearly wrong – a projection of the false concepts of 19th century science. Motion, whether in the quantum leaps of elementary particles or the movements of galaxies, is a derivative phenomenon, arising from the nature of matter, with its particles and anti-particles and the "reactions" connected with them. Engels goes further. He sees motion as an all-inclusive basic phenomenon embracing the biological, psychological and social realms: "for motion in nature and human history and for the motion of thought." But we have to examine the units of matter, mind and society – particles, neurons, social classes – in order to perceive the reasons for each of them acting as it does. When we do, it becomes apparent that there is no general "motion" common to them all. Motion in "human history" is not at all the same kind of thing as motion in "nature;" and "the motion of thought" is different from them both. True there is activity in all of them but this arises from the nature of the units involved. To consider it all as generalized "motion" obscures the true picture. There is the motion of the bodies of the universe and there is a motion of electrons in the neurons in the brain; but the first is the result of gravity, the second of the interactions of negatively and positively charged particles. "Motion in human history" arises from interpenetrating opposites of a social nature which have but a remote connection with those of matter. The changes involved in the evolutionary transitions from one form of matter or society to another cannot simply be seen as "the transition of one form of motion to another." To do so is to elevate a derivative characteristic to a primary – indeed deitylike – initiating position.[46]

In the days before science began to investigate the processes of matter and the universe, "motion" seemed to be a basic entity, and had, as we have seen, been a major concern of philosophy from the days of Aristotle's Prime Mover and before. If there was motion, the theologians (such as Aquinas) argued, it must have been implanted in matter by God. This the materialists from the early Greeks to the 18th century French disputed, arguing that motion was inherent in matter. Hence, there was no need to presume a God. Engels wrote under the influence of these philosophical concepts. But today motion is no longer a central philosophical question – at least for materialists (or physicists). Engels, who used scientific criteria whenever they were available, would certainly have accepted the findings of the 20th century in these as in other regards. Indeed, he noted that there would be such advances and that materialist philosophy would have to change in accordance with them.

The development of science in the present century has given us new insights into the specifics behind dialectical process, and these should replace philosophical disputation. "Contradiction," which also originated as a philosophical abstraction, ultimately reflects the fact that matter consists of opposite particles. That it should produce qualitative change by quantitative process is not a philosophical "mystery." When we begin with opposite entities, such as these particles, there is no way that they can produce new phenomena except by multiplying (or reducing) and combining. This is also exemplified in the workings of a computer, which begins with a simple positive and negative or plus and minus, and from them, and them alone, makes complex patterns upon patterns.

The "Categories" Fantasy

In his comments on dialectical thinking, Engels stressed interlocking fluidity in opposition to mechanistic abstractionism (as exemplified by Kant's system of categories). We must elevate our thinking above the rigidities of "true and false, good and bad, identical and different, necessary and accidental," for these "antitheses have only a relative validity." Engels developed his views further in a striking passage in *Socialism, Utopian and Scientific*:

To the metaphysician, things and their mental reflexes, ideas, are isolated, are to be considered one after the other and apart from each other, as objects of investigation fixed, rigid, given once for all. He thinks in absolutely irreconcilable antitheses. "His communication is 'yea, yea; nay, nay!'; for whatsoever is more than these cometh of evil." For him a thing either exists or does not exist; a thing cannot at the same time be itself and something else. Positive and negative absolutely exclude one another; cause and effect stand in a rigid antithesis one to the other.

At first sight this mode of thinking seems to us very luminous, because it is that of so-called sound common sense. Only sound common sense, respectable fellow that he is in the homely realm of his own four walls, has very wonderful adventures directly he ventures out into the wide world of research. And the metaphysical mode of thought, justifiable and necessary as it is in a number of domains whose extent varies according to the nature of the particular object of investigation, sooner or later reaches a limit, beyond which it becomes one-sided, restricted, abstract, lost in insoluble contradictions.[47]

By "metaphysical" here Engels does not mean only abstractionist in general but also, as the context indicates, categorical thinking of the kind introduced by Kant and expressed in the "formal logic" whose limits he elsewhere described. This "mode of thought," he indicates, does have a limited validity. Within artificially restricted areas, matters do appear to be black and white. Here simple "cause and effect" logic can apply. But if we pursue things into their interconnections, then "contradiction" and fluidity appear. A "common sense" approach to the facts of nature or society as revealed by scientific "research" cannot reflect the complexity of the reality. Unable to comprehend them we will be forced to take refuge in the simplistic abstractions of "good and bad, identical and different" and so on.

These limitations, Engels implies, permeate bourgeois thinking. Marx, as in his comments on Proudhon in 1846–1847, had come to similar conclusions: "The petty bourgeois is composed of on-the-one-hand and on-the-other-hand;" "instead of conceiving the economic categories as theoretical expressions of historical relations of production ... he garbles them into pre-existing, eternal ideas." These ideas are then imposed upon the realities. Some of these ideas involve "contradictions" (e.g., true and false). "Clever" people like Proudhon can "play with" them and produce "brilliant paradoxes" – "brilliant" but wrong; and sometimes, Marx sardonically notes, reactionary:

> Slavery [according to Proudhon] is an economic category like any other. Thus it also has its two sides. Let us leave alone the bad side and talk about the good side of slavery.

That bourgeois thinking in general is permeated by the kind of rigid abstractionism Marx and Engels described is hidden from us by its very universality. We simply get used to it as "normal" thinking. Let us take a passage not from a book on history or philosophy but, as more widely typical, one from a newspaper "think piece," Paul Johnson's "Colonialism is Back – and Not a Moment too Soon," *New York Times Magazine,* April 18, 1993:

> The Greeks, who invented colonialism, founded city-colonies to spread their civilization. The Romans, who inherited the Greek empire, did the same. ... From the Renaissance through to the early years of the 20th century, first the European powers and the Russian and the United States competed for colonies. ...

"Colonialism" to Johnson, then, is a "fixed" category, winging its way through history, ever unchanging, essentially the same in ancient Athens as in 20th century United States. Marx's comment on Proudhon in 1847 applies directly to the *Times'* article of 1993: "Instead of conceiving of the economic categories as theoretical expressions of historical relations of production . . . he garbles them into pre-existing, eternal ideas." And in so doing both he and Johnson obscure the historical reality. To speak of colonialism as an invention is, to say the least, crudely mechanistic, hiding the fact that social phenomena develop out of social forces. To speak of Greek colonialism as spreading "civilization" hides the hideous reality of a slave state bent on oppression and slaughter. For instance when, typically, the Greek "colonialists" conquered Melos, they killed the men and sold the women and children into slavery. The Romans, on a much larger scale than the Greek, slaughtered and enslaved by the tens of millions.

The "eternal idea," "colonialism" next touched down on Earth in the "Renaissance" (along with Shakespeare and Leonardo). What happened in the intervening ages? Did nobody happen to think of "reinventing" it for a thousand years? Or did the feudal states of that period not generate enough commercialism to sustain colonial conquest?

The cultural phenomenon known as the Renaissance arose on the back of a commercial revival, based first, for Italy and Spain, on the Mediterranean and, for Britain, on the North Sea and Baltic, and then for Spain and Britain, on the opening of the Atlantic for the brutal conquest of the Americas. As this commercialism was based on paid labor, it had a potential for development that a slave society, which stifled technological advance, did not have. As this commercial capitalism developed in the 19th century into industrial capitalism, it led to an even more extensive imperialist drive, especially when it assumed monopolist form in the 20th century. So, clearly all was not just "colonialism," popping up here and there in history, but different forms of imperialism arising from different socio-economic roots.

That a leading capitalist-world newspaper could, in 1993, publish this kind of simplistic, indeed mystical, rubbish reveals the cultural superficiality of the whole civilization. And behind all the obscurantism lies a threatening message: it is now "our" turn to spread our "civilization" throughout the world.

Although Marx attacked the "categories" of Proudhon as false reflections of "eternal" abstractions, he did not reject the idea of categories as such. Nor did Engels. What, then, is the place of categories in Marxist thinking?

The assertion of the importance of "categories" for philosophy was first made by Aristotle (perhaps with roots in Hinduism). Aristotle introduced ten categories: substance, quantity, quality, relation, place, time, position, state, action, passivity. The concept of categories was reintroduced by Kant, who listed twelve categories, including cause and effect, necessity and chance. Hegel accepted the concept but viewed the categories as fluid: quantity and quality "pass into each other." We should note, however, that – in the 2,000-odd years between Aristotle and Kant – although Bacon, Descartes, Spinoza, Leibnitz, Locke, Hume and others had to some degree used these forms of thought, they did so peripherally. Unlike Aristotle, Kant or Hegel, they saw no need for a system of categories in philosophy. Engels accepted Hegel's general concept of fluid categories.[48]

When that "respectable fellow," "common sense," ventures out into the "wide world" of nature as revealed by science, he finds that cause and effect "run into each other." Cause becomes effect, effect becomes cause in spiraling interaction. So, too, with basis and consequence, appearance and essence, identity and difference: "The plant, the animal, every cell is at every moment of its life identical with itself and yet becoming distinct from itself. . . . " And form and content: "The differentiation of form (the cell) determines differentiation of substance into muscle, skin, bone, epithelium, etc., and the differentiation of material in turn determines difference of form."[49]

The blending of chance and necessity is exhibited in social as well as natural evolution:

> Men make their history themselves, but not as yet with a collective will or according to a collective plan or even in a definitely defined, given society. Their efforts clash, and for that very reason all such societies are governed by necessity, which is supplemented by and appears under the forms of accident. The necessity which here asserts itself amidst all accident is again ultimately economic necessity.[50]

Some scientists believe they can do without basic generalizations. Others accept the generalizations provided by traditional philosophy, "categories" that are constructed by logical abstractionism without regard for science and are often theological in implication:

> Natural scientists believe that they free themselves from philosophy by ignoring it or abusing it. They cannot, however, make any headway without thought, and for thought they need thought determinations. But they take these categories unreflectingly from the common con-

sciousness of so-called educated persons, which is dominated by the relics of long obsolete philosophies . . . [51]

By at least the early 1970's Soviet philosophers had created a "system" of categories that they considered basic to dialectical materialism. Under "philosophical categories," the (collective) authors of *The Fundamentals of Marxist-Leninist Philosophy* list: "matter, motion, space, time, the finite and the infinite, consciousness, quantity, quality, measure and contradiction." In all they listed 24 categories. The *Great Soviet Encyclopedia* of the same period noted that philosophical categories "include singular, particular, general, part, whole, form, content, essence, phenomenon, law, necessity, contingency, possibility, reality, quantity, quality and measure among others." And these are only the philosophical ones. There are systems and sub-systems of categories in every field: history, economics, aesthetics, the natural science. The *Fundamentals* authors listed 16 entities with at least a semi-category status, including "stability and variability, equality and inequality, similarity and dissimilarity, identity and non-identity, imitability and non-imitability." Now it is certainly true that nature acts dialectically – by the fusion of opposite entities – and this is reflected in the processes uncovered by science. But this is not what the authors of the *Fundamentals* are saying. They propose a system of categories that stand above science and by which science is to be judged:

> The categories of dialectics differ from the general concepts of the specialised sciences, however, in that the latter are applicable only to a certain sphere of thinking, while philosophical categories, as methodological principles, permeate the whole issue of scientific thought, all fields of knowledge.

So, too, the *Encyclopedia:* "Categories form the basic intellectual means for the philosophical cognition of being and of the results of its concrete scientific and artistic reflection." The *Fundamentals* authors conclude: "A person must master the categories in the course of his individual development in order to possess capacity for theoretical thought." Marx, Engels and Lenin, of course, not only did not "master" this "system" of categories, they had never heard of it.[52]

This massive abstractionist structure, which one can only call "Kantianism run mad," dominated Soviet thought and the educational system for at least two decades, an early intellectual reflection of the social (class) forces that later attempted to undermine socialism.

Although Marx speaks of categories in various works, he does not present them as a unique ideological entity but as common generalizations. As examples of economic categories, he cites monopoly and competition: "Monopoly [according to Proudhon] is a good thing because it is an economic category ... Competition is a good thing because it is also an economic category."

Such things as monopoly and competition, however, are not specifically categories as such but simply general concepts: "Thus these ideas, these categories, are as little eternal as the relations they express. They are historical and transitory products."[53]

To Marx, then, categories are simply "ideas" along with other ideas. He does not anywhere subscribe to the concept of a system of categories or of thinking within the confines of such a system. In fact, he was suspicious of basically categorical thinking, ridiculing

> the metaphysicians who, in making these abstractions, think they are making analyses, and who, the more they detach themselves from things, imagine themselves to be getting all the nearer to their core.

And, he attacks Hegel's advocacy of fluid categories:

> Just as by dint of abstraction we have transformed everything into a logical category, so one has only to make an abstraction of every characteristic distinctive of different movements to attain movement in its abstract condition—purely formal movement, the purely logical formula of movement. If one finds in logical categories the substance of all things, one imagines one has found in the logical formula of movement the absolute method, which not only explains all things, but also implies the movement of things.[54]

The main categories that Engels discusses are cause and effect, identity and difference, form and content, chance and necessity. In considering them, we have to note at the outset that they are, as Engels said of chance and necessity, "thought determinations." At times, however, Engels seems to regard them as objective phenomena.

As with "law," to see "cause" as inherent in nature is idealist. What we have are interactions of the parts of matter, some of them directly, some indirectly, oppositional. There is neither cause nor effect, only the things and their interactions.

Similarly with "identity and difference." All atoms have their general structure in common but differ in their constituent number of protons, neutrons and electrons. So, too, in biology. Cells have both individual characteristics (liver cells, brain cells), but also basic characteristics in common. In the total animal we perceive both sexual differences but also, for each species, a basic area of sameness.

The situation in regard to "chance" and "necessity" is similar to that we noted for "law or "cause," namely that certain complexes of interactions can lead only to certain other complexes, as in an embryo, complexes built by evolution. Chance seems to reflect interactions in which a number of new complexes is possible. The concept "form and content" basically reflects the fact that when atoms, particles, molecules or cells join together they necessarily have some shape and size, as determined by their number, nature and arrangements. In clumped matter, size and shape are determined by gravitational as well as electromagnetic and nuclear reactions. In biological matter, form is determined by the interactions of its units at various levels, from macromolecules to cells to bodily organs.

If we thus pursue the dependence of the categories on the nature of matter (and, in another context, of society also) we have to ask what do "categories" – including materialist ones – have in common that would distinguish them as a separate form of generalization? The answer is, nothing. They are simply general ideas along with other general ideas reflecting aspects of reality. To link them together as a unique philosophical entity gives the impression that they have some special, perhaps "higher," status than other general ideas; even, in some contexts, that they are basic determinants. All in all, it seems best to drop the term "category" and to speak only of general ideas, allowing the context to indicate the character of the phenomena they reflect. The term category inevitably retains something of its original idealist significance of fitting reality into set frames of thought (Kant) or into a mystical Godhead (Hegel). Once we begin with the concept of categories, we are inevitably led to their systematized listing. And here problems arise. Which general concepts are to be considered as categories and which are not? If we include all general concepts the list is virtually endless. So, too, the debate on what to include and what to exclude.

There is another problem also. Are Engels' "fluid categories" really categories? Is, for instance, an interblending cause and effect, in which cause becomes effect and effect cause as they reflect the interactions of matter (or society) really a category? Certainly not in the basic meaning of the word. So, too, with identity and difference, necessity and chance, and so on. To call them categories confounds their relationship to the endlessly interpenetrative reality they mirror, giving it a rigid and abstractionist tinge (as Marx implied in his attack on Hegel's fluid categories).

Differences Between Marx and Engels

As we have seen, there are differences between the writings on dialectical materialism by Marx and by Engels. If Marx and not Engels had developed the subject, he would clearly have handled matters somewhat differently. He would have placed the interpenetration of opposites more consistently at the center of things and put more emphasis on the "critical and revolutionary" nature of "scientific dialectics," on materialism as the philosophical base, on practice, on the interactive role of the senses and activity in the knowing process. He would have taken a more sharply critical attitude towards Hegel and he would not have written passages in which interaction and not contradiction could appear to be the essence of dialectical process. He would not have elevated "dialectics" to a state of unique eminence in the materialist world-view or in materialist ways of thought.

How can we reconcile these differences with the fact that Engels, as he tells us, "read the whole manuscript [of *Anti-Dühring*] to him [Marx] before it was printed," and Marx added a chapter (on economics) to it? The answer is that Engels' divergences are not so clear in the philosophical section of *Anti-Dühring* as they are in *Feuerbach,* which was written after Marx's death, and in the *Dialectics of Nature* manuscripts, which Marx apparently did not see. The philosophy section of *Anti-Dühring* is largely concerned with science – Time and Space; Cosmogony, Physics, Chemistry: The Organic World – and Morality and Law. Only the two final chapters are on dialectics: Quantity and Quality: Negation of the Negation. And in these chapters there seems little that Marx would not have agreed with. If he had had strong objections, he would certainly have voiced them, but he may have hesitated to discuss what seemed to be secondary differences. After all, it was Engels' book and not his.

Engels was aware that Marx had a deeper grasp of their world-view than he did. "Marx," he wrote, "was a genius; we others were at best talented." There were, of course, no "others" on the same level as Engels. And it was not only Marx's inherent intellectual superiority, his potential "genius," but his background that made the difference. Marx had been in bitter political struggles from an early age, had been persecuted, exiled and imprisoned, whereas Engels had led a comparatively quiet life. But whatever the reason, it is apparent that, as Engels also noted, Marx "saw further and took a wider and quicker view." Behind this view was his immersion, physical and mental, in the class struggle of the proletariat. In his political writings, for instance *The Civil War in France,* there is a revolutionary fierceness that contrasts with Engels more academic treatment of the same subject (the Paris Commune).[55]

We should not, of course, overstate the differences between the two, especially as there has arisen in recent decades an unsupportable theory that they were in fundamental disagreement and that Marx took little or no interest in dialectical materialism or science. On the other hand we cannot accept the view, spread in Soviet and other Marxist circles, of an absolute harmony between the two. This absolutist argument, in fact, was used to promote Soviet idealist tendencies that are seen as having roots in Engels.

The concept of a complete agreement between Marx and Engels probably stems originally from a loose reading of Lenin, who in his *Granat* encyclopedia article on Marx, in *State and Revolution* and other works, uses passages from Engels interchangeably with those of Marx (as we shall note). Lenin emphasised the work of Engels as illustrating certain aspects of Marxism that were being overlooked or distorted at the time by the social-democratic revisionists. But he was aware, as he indicates in *On Dialectics,* of deficiencies in Engels. It is time now to take a more precise view and see in what ways Marx did "see further," what Engels' actual deficiencies were and what their effects have been. Whatever deficiencies there are, however, the fact remains that Engels' studies, especially of science, gave materialism new scope and depth. *Anti-Dühring, Feuerbach* and *Dialectics of Nature* revealed in detail for the first time the working of the dialectical process in nature and demonstrated the necessity for dialectical reasoning in science.

The "Dialectical and Historical Materialism" Fallacy

Marxism, both social and philosophical, presents general outlooks that are largely outside the scope of exploitive-class thinking and so are confusedly dismissed by bourgeois commentators. This is clearly seen when Marx's social perspectives run directly counter to the dominant views of corporate capitalist society, as in his exploration of the exploitation of labor. But it is true also in regard to dialectical materialism, whose anti-religious implications evokes religious bias, directly or indirectly. Hence it is often represented as a mosaic of doctrinaire abstractions rather than – as it is – a necessarily complex reflection of the workings of nature and society. Bourgeois thinkers can, of course reason well within their mainly static framework, and have, in fact, achieved remarkable, if fragmented, insights into reality, social and natural, but they lack the materialist base that could keep them from mixing disparate elements, and the "scientific" dialectic that could enable them to perceive underlying interconnections. As a simple example we might compare Lenin's analysis of the new Soviet state with that of the leading bourgeois thinkers of the time. H.G. Wells, for instance, in his *The Outline of History,* saw the early U.S.S.R. as essentially embodying a struggle between "Westernizers" and "Easternizers" (who were "primitive and mystical"). The remark is typical. In general the "great thinkers" of "Western Civilization" had little comprehension of what had happened. On the other hand, in *Letters from Afar* and other writings, Lenin indicated the basic (class) nature of the situation even before it had unfolded and correctly charted its future (including the function of the Soviets).[56]

Although dialectical materialism is widely regarded by philosophers and others in capitalist nations as dogma, it would be hard to conceive of less dogmatic works than Engels' fact-filled *Dialectics of Nature* or, as we shall see, Lenin's *Philosophical Notebooks,* with its spirit of analytical inquiry. The charge, however, has received some support from the semi-dogmatic method that Stalin developed and that was expanded by later Soviet thinkers. There was, however, no basis in Stalin for the later categories madness.

In Soviet textbooks, following Stalin's famed essay, *Dialectical and Historical Materialism,* Marxism was presented as falling into two parts, "dialectical materialism" and "historical materialism," and so in Marxist parties throughout the world. "Historical materialism" is used to designate society in general. This, however, is clearly incorrect. All the aspects of society cannot be subsumed under "history" or "historical."

Marx apparently did not use the term "historical materialism." Engels used it at least once, in *Socialism: Utopian and Scientific* (1880). His usual term, in *Anti-Dühring* and *Feuerbach,* is "the materialist conception of history"; namely "that view of the course of history" that sees "economic development" and the class struggle as providing the "moving power of all important historical events." "Historical materialism" is a kind of shorthand for this concept. Neither this shorthand term nor the concept it stands for, however, encompass social life, including the major areas of family, community and personal relations. True, these areas in some ways respond to historical evolutionary factors but they are rooted primarily not in these but in socio-biological process and have their own essence and developmental rhythms. Nor does "historical materialism" properly cover culture and the arts – literature, music, painting, dance and so on. These, too, are changed by historical change but their roots are in social life and their essence is in the nature of people. Literature is literature in every age and society. Nor does "historical materialism" properly include science, natural or social, for although science develops – or declines – in response to socio-economic forces, it is a cultural phenomenon with a special character rising from its base in practice. Finally, human psychology and thought as such are not basically controlled by the historical process even though they respond to it. The ideas of everyday life are primarily part of general "social consciousness," not of the "religious, philosophical" and other ideological "forms" that respond directly to political or economic change.[57]

However, although "historical materialism" does not cover these other aspects of society, "Marxism" – which embraces a general science of society – does, for it examines all its aspects, including psychology and the arts. The division, then, should not be between dialectical materialism and historical materialism but between Marxist philosophy and Marxist social science. If "historical materialism" is used at all it should be used in the restricted sense indicated by Engels. Nor should it be used in juxtaposition with "dialectical materialism," a juxtaposition which implies that all reality falls into two master categories. The juxtaposition is not to be found in Marx, Engels or Lenin but first achieved prominence in Stalin. In short, the formula "dialectical and historical materialism" should be dropped.[58]

The Limits of "Marxism-Leninism"

Another problem is inherent in the term "Marxism-Leninism." In some Soviet and other studies, Marxism-Leninism is used to include and indeed supercede, Marxism, as in the title *Fundamentals of Marxist-Leninist Philosophy.* This usage is patently wrong. Lenin examined in particular three major historical phenomena: monopoly capitalism, the proletarian revolution, and socialism. His analysis of all three carried Marxism to a new level, theoretical and practical. Beyond this, he added to Marxist understanding in every field. In fact, we can hardly turn a page of his works without finding concepts that add depth or scope to Marxism. Nevertheless Marxism has not been transformed into or become part of Marxism-Leninism. Marxism-Leninism refers to Lenin's specific developments of Marxism and to his general interpretation of Marxism, which was in essence that of Marx himself (as Lenin cleared the air of revisionism and dogmatism). The *Fundamentals* and Soviet theoretical works in general, however, not only included "historical materialism" under Marxism-Leninism but also dialectical materialism, which is confusion worse confounded. Although Lenin greatly expanded the horizons of dialectical materialism, especially by his examinations of science, epistemology and dialectical method, this does not make dialectical materialism a subsidiary part of Marxism-Leninism. The *Fundamentals'* view reflects that Russian chauvinism that more and more permeated Russian society in the decades before the political assault by the capitalist-oriented forces.

The authors of the *Fundamentals* wrote that Marx and Engels "created historical materialism by extending philosophical materialism and materialistically revised dialectics to the interpretation of society." Further: "When applied to society, the dialectical method becomes concrete." The implication that Marx and Engels first invented a philosophy and then extended it to society is obviously wrong. Marxism arose from the practical experiences and observations of Marx and Engels in a special "revolutionary epoch." Marxist "philosophical materialism" arose from their science-based and materialist analysis of the relations of humanity and nature, an analysis sharpened by their social views and activism. The further implication that "the dialectical method" becomes "concrete" only when "applied to society" really denies its philosophical validity. It is concrete in regard to both nature and society, and in neither case because it is "applied," but because it reflects reality. The authors of the *Fundamentals* were again echoing Stalin (and again without acknowledgment): "Historical materialism is the extension of the principles of dialectical materialism to the study of social life." Like some other formulations of Stalin, this one bears the Aristotelian imprint of his early theological training.[59]

Chapter Three

LENIN

Although Lenin's writings on philosophy were well known in the U.S.S.R. and other socialist countries they have been little known in the capitalist world. Even those who have read *Imperialism* or *State and Revolution* often know nothing of *Materialism and Empirio-Criticism* or the *Philosophical Notebooks*. Yet these philosophical writings make important advances in materialist thought, going, in some ways, beyond Marx and Engels.

Religion and Marxism

Marx and Engels argued that Protestantism had helped the European commercial bourgeoisie in its long battle against feudalism and the feudal church. With the triumph of the bourgeois revolution in Europe, however, Engels wrote in 1886, "Christianity entered its final stage" and became "incapable for the future of serving any progressive class as the ideological garb of its aspirations." Neither he nor Marx, however, advocated anti-religious planks in radical political platforms. When in 1891 the German Social Democratic Party advocated a strong anti-Church position, Engels suggested a compromise. Time enough to deal with religion after the seizure of power by the workers. In the meantime, one must oppose it on the intellectual level but not allow it to become a political issue to disrupt the "movement of the present" that would bring about "the future of that movement."[1]

Lenin, as the leader of a party advancing to the revolutionary seizure of power, and then, from 1917 to 1923, as the main architect of an emergent proletarian state, faced the problem of religion with greater urgency than had Marx and Engels. "Marxism," he wrote in 1908, "always regarded all modern religions and churches, and every kind of religious organization as instruments of that bourgeois reaction whose aim is to defend exploitation and stupefy the working class." But he opposed making atheistic propaganda part of the Party program: "We must not only admit into the Social-Democratic Party all those workers who still retain faith in God, we must redouble our efforts to recruit them. We are absolutely opposed to the slightest affront to these workers' religious convictions." He would go even further: "If a priest comes to us for common political work – if he conscientiously performs Party work, and does not oppose the Party programme, we can accept him into the ranks . . . " Moreover, to make a full-scale ideological attack on religion would be ineffective because religion arises from deep needs, created by the poverty, misery and oppression generated by capitalism:

> The deepest root of modern religion is embedded in the social oppression of the working masses, and in their apparently complete helplessness before the blind forces of capitalism, which every day and every hour inflicts a thousand times more· horrible suffering and torture upon rank and file working people than are caused by exceptional events such as war, earthquakes, etc. "Fear created the gods."[2]

Religion might be "the opium of the people" but, as Marx perceived, the people need the opium to deaden the horrors inflicted on them in a "soulless," exploitive society. While in the course of revolutionary struggle some workers might slough off their religious views and this would make them better Marxists, the most important thing, Lenin felt, was to advance the mass struggle against capitalist oppression, and here one has on the whole to accept people as they are. The struggle was also, of course, directed against feudal oppression, which in Czarist Russia created a wider social base for religion than did capitalism, and, moreover, had control of the Church. But Lenin was here concentrating on the situation of the industrial working class.[3]

Even after the Bolsheviks had obtained State power Lenin opposed a head-on political attack on religion but instead envisaged a long-term educational campaign (1922):

> It would be the biggest and worst mistake which a Marxist could make to think that the millions upon millions of people (particularly the peasants and artisans), who have been condemned by the whole of modern society to darkness, ignorance and prejudice, can emancipate themselves from this darkness solely along the straight line of purely Marxist education.

In developing a socialist order, the uprooting of religion must be approached cautiously and on a broad ideological front:

> These masses must be given the most diverse, atheistic propaganda stuff, they must be made acquainted with the facts from the most different spheres of life, they must be approached one way or another

so as to get them interested, rouse them from their religious torpor and stir them up from the most different angles and by the most different methods, etc.

A Marxist leader should make use of the "clever, vivacious and talented journalism of the old atheists of the eighteenth century" (presumably Diderot and Holbach, and, perhaps – extending "atheists" to include other anti-clericals – the satiric wit of Voltaire). Such "journalism" could be easily understood.[4]

As we read Lenin's comments on religion we become aware that he had the opportunity to actually bring into being projects that Marx and Engels could but look toward. Thus while they worked out a basically correct practical and theoretical position on religion, something that no radical thinkers, socialist or otherwise, had been able to do before, Lenin was able to take this position, develop it in the light of experience and apply it on a vast scale, seriously challenging, for the first time in history, the narcotic of mass superstition.

Although Marx, Engels and Lenin all felt that direct attacks on religion should be avoided because of its deep roots among the masses, they had no such reservations about reactionary philosophy. Such philosophy was not rooted in mass oppression but in the need to confuse progressive intellectuals and disorient the leadership of progressive movements. It must be met in head-on debate (as Engels did in regard to the new milk-and-water skepticism of the "agnostics").[5]

Idealism and the Mad Piano

After and even before the death of Engels in 1895, most leaders of the socialist movement ignored his writings on philosophy. Karl Kautsky, who was widely regarded as Engels' successor, felt that philosophy was irrelevant to Marxist theory, which he regarded as social doctrine and nothing more. The only major exception was Georgi Plekhanov, the outstanding intellectual in the Russian Social Democratic Party, who tried to develop Engels' views in a series of philosophical studies, and in 1891, as we saw, coined the term "dialectical materialism." Plekhanov examined the roots of Marxist philosophy in the French materialists and Feuerbach. He stressed especially the concepts of "leaps" in development:

Many people confuse dialectic with the doctrine of development; dialectic is, in fact, such a doctrine. However, it differs substantially from the vulgar "theory of evolution," which is completely based on the principle that neither Nature nor history proceeds in leaps and that all changes in the world take place by degrees. Hegel had already shown that, understood in such a way, the doctrine of development was unsound and ridiculous.[6]

As the comment suggests, Plekhanov was a revolutionary at a time when his colleagues in Western Europe were mainly parliamentary reformists. Russia, unlike Western Europe, had not yet undergone an anti-feudal revolution, even though capitalism had made considerable advances, but was governed by a basically feudal dictatorship that could only be overthrown by revolution.

Lenin, as a promising radical intellectual growing up in the Party when Plekhanov was its leading theorist, shared his interest in philosophy. In political exile in Siberia in 1899 (he was born in 1870) he became alarmed at the growth of "Neo-Kantianism" among Russian radicals and began to read in philosophy:

> And that is what I am studying now, having begun with D'Holbach and Helvetius, and I intend to pass on to Kant. I have got hold of the most important classics of philosophy but I have no Neo-Kantian books.[7]

Following the defeat of the 1905 Revolution, a number of radical intellectuals, including A.V. Lunacharsky, joined a fashionable swing to a neo-Kantian idealism then known in Europe as empirio-criticism. Largely under the influence of a prominent Austrian physicist, Ernst Mach (1838-1916), it was given a "scientific" cloak. Mach's *Analysis of Sensations* and other philosophical works, eclectically mingling physics and psychology, first expressed Humean skepticism (all we know are sensations) and then Kantian idealism (space and time are mental constructs). By 1908 at least four books attacking dialectical materialism had appeared in Russia, but what mostly disturbed Lenin were works by Bolshevik intellectuals, including a series of articles in a book titled *Outlines of Marxian Philosophy*. "Every article," Lenin wrote from Geneva to Maxim Gorky, the leading "Party" author (1868-1936), "made me simply furious. No, that is not Marxism. Our Empirio-Critics . . . are simply sinking into a bog." And he informed Gorky, who had himself taken a somewhat ambivalent stand, that he was going to write about these matters, Plekhanov having failed to do so:

> You must understand—and you will, of course—that once a Party man has become convinced that a certain doctrine is grossly fallacious and harmful, he is obliged to come out against it . . . Plekhanov, at bottom,

is entirely right in being against them, only he is unable or unwilling or too lazy to say so concretely, in detail, simply, without unnecessarily frightening his readers with philosophical nuances. And at all costs I shall say it in my own way.

By the fall of 1908 Lenin had finished the book – working first at the Bern Library in Switzerland and then at the British Museum – and the following year it was published in Moscow: *Materialism and Empirio-Criticism: Critical Comments on a Reactionary Philosophy.*[8]

It becomes clear early in his book that Lenin is exploring new territory. Marx and Engels had no need to concentrate on the particular brand of idealism, subjective idealism, that stemmed from Berkeley and was refined by Kant. Lenin, however, because of the emphasis then being placed upon it by Mach and others, has to do so – and does it in his "own way."

> For every scientist who has not been led astray by professional philosophy, as well as for every materialist, sensation is indeed the direct connection between consciousness and the external world; it is the transformation of the energy of external excitation into the fact of consciousness. This transformation has been, and is, observed by each of us a million times on every hand. The sophism of idealist philosophy consists in the fact that it regards sensation as being not the connection between consciousness and the external world, but a fence, a wall, separating consciousness from the external world—not an image of the external phenomenon corresponding to the sensation, but as the 'sole entity.'[9]

Mach, Lenin noted, defined physics as "the discovery of the laws of the connection of sensations." In physics then, we are not uncovering facts but only sensations; the facts – Kant's "thing in itself" – remaining forever unknowable.

Although Lenin, following Engels, here speaks of "image," he is clearly not, as has been charged, taking a simple mirror-image view of cognition. His concept of "the energy of external excitation" being "transformed" into "sensation" and sensation into "consciousness" implies an active process in line with Marx's linking of mind with "human sensuous activity." But he goes beyond Marx in placing the emphasis – doubtless reflecting the new electron physics – concretely on "the energy of external excitation." The total process he seems to regard as a manifestation of the transformation of energy (which Engels had singled out as a major discovery).

Lenin next considers Humean skepticism:

> By skepticism Hume means refusal to explain sensations as the effects of objects, spirit, etc. refusal to reduce perceptions to the external world, on the one hand, and to a deity or to an unknown spirit, on the other.

Far apart though they may seem, Berkeley and Hume have a base in common: they are both subjectivists, even though the one saw only mind and the other only sensation. For a refutation of both, Lenin turned to one of the most "vivacious" of the 18th century "atheists":

> In the "Conversation Between d'Alembert and Diderot," Diderot states his philosophical position thus: "... Suppose a piano to be endowed with the faculty of sensation and memory, tell me, would it not of its own accord repeat those airs which you have played on its keys? We are instruments endowed with sensation and memory. Our senses are so many keys upon which surrounding nature strikes and which often strike upon themselves.... There was [according to Berkeley's view] a moment of insanity when the sentient piano imagined that it was the only piano in the world, and that the whole harmony of the universe took place within it."

In short, Lenin suggests, both subjective idealism and Humean skepticism end in solipsism (only "I" exist), a doctrine that he considered *ipso facto* absurd. Those who, like Mach, thought they were cleverly rising above theological crudities by their refined idealist and skeptical views were still boxing themselves into a solipsistic absurdity.[10]

It was partly in an attempt to avoid the philosophical chaos inherent in Humean skepticism that Kant erected his scaffolding of built-in mind categories and the ever-unknowable "thing in itself." Engels, countering this view, argued that as the reality of nature was disclosed by science and was harnessed by technology, Kant's "thing in itself" became a "thing for us," something people could understand and work with. The basic criterion was not thought but practice: "human action had solved the difficulty long before human ingenuity invented it." Lenin agreed, but he emphasized that practice was not an absolute criterion: "Of course, we must not forget that the criterion of practice can never, in the nature of things, either confirm or refute any human idea completely." Nevertheless, science does provide a general and verifiable correspondence between our ideas and objective phenomena:

> A man in a dark room may discern objects dimly, but if he does not stumble over the furniture and does not walk into a looking-glass instead of through a door, it means that he sees some things correctly. There

is no need, therefore, either to renounce the claim to penetrate below the surface of nature, or to claim that we have already fully unveiled the mystery of the world around us.

This, in spite of the deceptive simplicity of Lenin's analogy, has deep implications, going beyond Engels, in examining the limitations of "practice" as a path to truth, but also affirming its basic validity. "Practice," here specifically the practice of scientific exploration, gives us, Lenin suggests, only the basis for varifying reality. It does not "fully" unveil the world. All we can hope for are approximations, of a greater or lesser degree of accuracy. These, however, are – the implication, here and elsewhere, runs – close enough to enable us to change reality, social and natural.[11]

If human knowledge is not "absolute," is it not simply "relative," as the neo-Kantians had been arguing? Indeed, on the basis of the Marxist argument itself that ideas are the product of specific social conditions, must not all knowledge be relative? Lenin answers as follows:

When and under what circumstances we reached, in our knowledge of the essential nature of things, the discovery of alizarin [a dye] in coal tar or the discovery of electrons in the atom is historically conditional; but that every such discovery is an advance of "absolutely objective knowledge" is unconditional. In a word, every ideology is historically conditional, but it is unconditionally true that to every scientific ideology (as distinct, for instance, from religious ideology) there corresponds an objective truth, absolute nature. You will say that this distinction between relative and absolute truth is indefinite. And I shall reply: it is sufficiently "indefinite" to prevent science from becoming a dogma in the bad sense of the term, from becoming something dead, frozen, ossified; but at the same time it is sufficiently "definite" to enable us to dissociate ourselves in the most emphatic and irrevocable manner from fideism [religion] and agnosticism, from philosophical idealism and the sophistry of the followers of Hume and Kant.

If a scientific concept, such as that of the electron, is upheld by experimentation it must have a basic correspondence to objective reality regardless of the social circumstances that produced the concept; and the more widely it can be used to produce predictable results, the greater must be the correspondence. The fact that such truths are unearthed because of particular social circumstances does not mean that they are any the less true. On the other hand we cannot claim for any of them an "absolute" correspondence with the physical reality. Nor are they simply "relative" – with the implication of no possibility of verification. As with the man in the dim room analogy, Lenin argues for a general correspondence only, one close enough to the reality to enable us to create change on its basis. This scientific approach to truth, whatever its limitations, is on a different epistemological plane to that of idealism or skepticism – whose "truths" are merely delusions.[12] As I suggested in discussing Engels' comment on these problems, it seems best to drop the term "absolute truth" for it tends to perpetuate the idealist Mind-to-mind revelatory communion in which it is rooted. Similarly with such phrases as "absolutely objective knowledge" and "objective truth." Neither knowledge nor truth are objective. The words indicate the mind's perception of reality. To equate "objective truth" with "absolute nature" makes matters worse. Nature is neither relative nor absolute. Lenin's terms mix subjective and objective phenomena, something he is generally careful to avoid. The problem is that he is dealing with a complex subject. Although objective and subjective are different in their individual essences, there is an area of overlap between them. Hence difficulties in nomenclature tend to arise. Lenin would have done better here to have based himself on his external "excitation" to "sensation" to "consciousness" model and avoided ambiguous usage of "knowledge" or "truth."

Today the situation has been complicated by the discovery that not only nature but the human mind is very different from what it appears to be. The mind does view reality in terms of constructs; not, however, the unreal ones of Kantian categories but real ones imposed by the mind's evolutionary history. The specifics of this problem we can better discuss in considering modern science.

It might seem that scientists or philosophers acquainted with science could themselves come to materialist conclusions. But although some of them do (to varying degrees) they are on the whole, Lenin argues, hampered by their class position. The situation in philosophy, despite its rarified abstractionism, is essentially the same as in other fields:

> Not a single one of these professors, who are capable of making very valuable contributions to the special fields of chemistry, history or physics, can be trusted one iota when it comes to philosophy. . . . Taken as a whole, the professors of economics are nothing but learned salesmen of the capitalist class, while the professors of philosophy are learned salesmen of the theologians.[13]

Although scientists in the capitalist world today seem by and large ready to challenge direct religious intrusions into their field, as indicated, for instance, in their opposition in the U.S. to the teaching of so-called "creationist science," there are still a sizeable number that proclaim idealist and sometimes even mystical views – as witness *The Omni Interview* (1984), interviews with contemporary scientists in various fields. Here we find Francis Crick, the co-discoverer with James Watson of the "double helix" replicating pattern of the nucleic acids, speculating that life on earth came from a space ship sent to implant it by some "creatures like ourselves" from a remote planet beyond the solar system, a "theory" which he labels "directed panspermia." Candace Pert, a prominent neuroscientist, after giving a materialist account of the brain and its neurotransmitters astoundingly comments: "the brain may not be necessary to consciousness. Consciousness may be projected to different places." And she affirms faith in God and life after death. John Lilly, an animal psychologist, comments: "In this self-transcendence one can experience bliss while performing God's work." And: "Newborns are connected to the divine; war is the result of our programmed disconnection from divine sources." Roger Sperry, a neurologist famous for his explorations of the brain's hemispheres, laments "the rejection of institutional religion" by "the Marxists" and "secular humanists." Brian Josephson, a Nobel Prize winner in the field of superconductivity, regards "inspiration" as the basis of scientific discovery, and adds: "inspiration may be the universal intelligence communicating with your brain." Freeman Dyson, a physicist at the Princeton Institute for Advanced Studies, comments on human evolution: "I don't think you can explain that by natural selection at all. It's just a miracle." And "it almost seems as if the universe in some sense must have known that we were coming." Clearly in a society in which the irrational and the supernatural are daily propagated, sometimes blatantly, sometimes subtly, scientists are not immune.[14]

The "Inexhaustible" Electron

Between Engels' death in 1895 and 1908, when Lenin wrote *Materialism and Empirio-Criticism,* science laid the basis for the elementary particle physics of the present century, thus enabling Lenin to go beyond Engels in developing the materialist concept of matter.

From the mid-19th century, experiments had been conducted on the effects of an electric current passing through a vacuum tube. In 1895 the British scientist Joseph John Thomson ascertained, by magnetically and electrically deflecting this current, that its particles ("electrons") were negatively charged and calculated that they were some 1,800 times less massive than the atom. In the same year, the German physicist, Wilhelm Roentgen, found to his surprise that when electrons struck metal they generated a form of radiation (X-rays) that could penetrate matter (the general phenomenon of electrons producing photons). They also caused phosphorescent substances to glow. The following year, Henri Becquerel, in France, in considering the effect of these rays on phosphorescence happened by chance to pick uranium nitrate for one of his experiments. As the "rays" coming from the uranium nitrate continued to glow after the electricity was turned off – as the X-rays did not – he suggested that they were not caused by the electrons striking the uranium nitrate but were somehow generated within it. In 1898 Marie and Pierre Curie extracted a substance from pitchblende that was even more "radioactive" than uranium, which they called radium. In 1899, Ernest Rutherford, at McGill University in Montreal, broke down the radiation emerging from radium into three "rays," which he called alpha, beta and gamma rays. The beta rays were soon found – by the Curies and Becquerel among others – to be simply electrons travelling at a high speed. In 1906 Rutherford ascertained that the alpha rays were positively charged particles and later calculated that they had a mass 7,000 times that of the electron. When he shot these particles through a thin glass into a chamber from which they could not escape he found that the chamber contained helium. The alpha particles, it thus appeared, were helium atoms (actually they were the nuclei only). The relation of these various particles to each other, however, was not clear, and it was not until 1911 that Rutherford suggested a model of the atom with a positively charged nucleus of large mass with negatively charged electrons revolving around it. In the same years Einstein began his undermining of the Newtonian universe, bringing out in 1905 both his "special," i.e., limited, theory of relativity – that embraced both the movement of bodies and the equivalence of mass and energy – and his quantum (photon) theory of light (following Max Planck's general quantum theory of 1900).

In discussing the philosophical implications of some of these discoveries, Lenin made it clear at the outset that he was doing just that and nothing more:

> It goes without saying that in examining the connection between one of the schools of modern physicists and the rebirth of philosophical idealism, it is far from being our intention to deal with specific physical theories. What interests us exclusively is the epistemological conclusions that follow from certain definite propositions and generally known discoveries.[15]

By not examining "specific physical theories" Lenin meant that as he was not a physicist he was not qualified to examine specialized areas. He was primarily interested in seeing what new aspects of matter science had revealed and how this affected materialist views. For instance, in considering the discoveries of Thomson and Rutherford, he began to move away from the "chemical" analysis of the "unchangeable" matter of 19th century science:

> Each day it becomes more probable that chemical affinity may be reduced to electrical processes. The indestructible and non-disintegrable elements of chemistry, whose number continues to grow as though in derision of the unity of the world, prove to be destructible and disintegrable. The element radium has been converted into the element helium.

Lenin, then, early (1908) grasped in a general way two related concepts of 20th century science, namely that molecular ("chemical") interactions are ultimately a response to "electrical processes" – essentially negatively and positively charged particles – and that the atom itself can be broken down: "the disintegration of particles of matter which hitherto had been accounted nondisintegrable." In the same paragraph as that quoted above, he refers to "X-rays, Becquerel rays, radium." He was aware of the discovery of the electron – "corpuscles charged with negative electricity" – , its velocity – "270,000 kilometres per second" – and its size relative to the atom as a whole: "The electron is to the atom as a full stop in this book is to the size of a building 200 feet long, 100 feet broad, and 50 feet high." In his footnotes he lists, among other works, J.J. Thomson's *The Corpuscular Theory of Matter* and Oliver Lodge's *Electrons.* He did not then know of the existence of the proton or the role of the electron within the atom for these were not discovered or theorized until 1911. His interest in science, however, continued beyond *Materialism and Empirio-Criticism.* A notebook of 1909 lists a book on physics by Max Planck and others. In 1914 he quoted Engels – "in the last analysis, nature proceeds dialectically" – and comments that this was "written before the discovery of radium, electrons, the transmutation of elements." In 1922 Lenin commented that while Einstein's relativity theories had been interpreted idealistically, Einstein himself "makes no attacks on the foundations of materialism."[16]

The new discoveries in physics had been immediately utilized by neo-Kantian idealists and skeptics to proclaim, as Lenin noted, that "matter has disappeared." Lenin replied that what modern physics had demonstrated was not the disappearance of matter but simply the existence of hitherto unknown aspects of matter:

"Matter disappears" means that the limit within which we have hitherto known matter disappears and that our knowledge is penetrating deeper; properties of matter are likewise disappearing which formerly seemed absolute, immutable, and primary (impenetrability, inertia, mass, etc.) and which are now revealed to be relative and characteristic only of certain states of matter. . . .

The world is matter in motion, we reply, and the laws of its motion are reflected by mechanics in the case of moderate velocities and by the electromagnetic theory in the case of great velocities.

Matter, then, is not "immutable" but is composed of various "states" unknown to Marx and Engels. What these "states" are is becoming clearer today and it appears that what basically distinguishes them is not variation in "velocity" but combinations of particles and the "interractions" that are connected with them: nuclear, electromagnetic, gravitational and "weak" (which causes atomic decay). Variation in velocity is, like "motion," a by-product of these phenomena. Neither is a basic initiating entity.[17]

Scientists should, Lenin argues, begin to distinguish between the "old" materialism – with its "atomistic-mechanical conception of nature" – and dialectical materialism with its fluid-conflictive outlook:

The destructibility of the atom, its inexhaustibility, the mutability of all forms of matter and of its motion, have always been the stronghold of dialectical materialism. All boundaries in nature are conditional, relative, movable, and express the gradual approximation of our mind towards knowledge of matter.

Lenin was probably thinking primarily of the fact that the dialectical materialist view of matter encompasses its "mutability" in general, a mutability that, it could be argued, entailed the potential "destructibility of the atom." But he may also have known that Engels had expressed specific views on these matters, reflecting a controversy on them at the time. Chemists, Engels noted in 1877, were inclined to believe that the atoms were not "the smallest known particles of matter" and "the majority of physicists" believed that the "ether" consisted of particles smaller than atoms. The atom, he informed Marx, is no longer considered "indivisible."[18]

Unlike the old "mechanical" materialism – then still practiced by the followers of Ernst Haeckel and others – dialectical materialism, Lenin contends, can not only encompass the discoveries of Thomson and Rutherford and the Curies, it is the philosophy required by the new science if it is to fully understand the significance of its findings.

Recent scientific advances provided increasing insights into the nature of matter and the universe:

> The "essence" of things, or "substance," is also relative; it expresses only the degree of profundity of man's knowledge of objects; and while yesterday the profundity of this knowledge did not go beyond the atom, and today does not go beyond the electron and ether, dialectical materialism insists on the temporary, relative, approximate character of all these milestones in the knowledge of nature gained by the progressing science of man. The electron is as inexhaustible as the atom, nature is infinite, but it infinitely exists.[19]

As in the "destructibility of the atom" passage, Lenin here follows Engels' view on scientific advance as the path to knowledge, revealing ever new aspects of "essence" and "substance." The atom passage also helps to elucidate his meaning in his often quoted but seldom examined comment here, that the "electron is as inexhaustible as the atom," for there he speaks of the atom's "destructibility," and "inexhaustibility." Hence by "inexhaustible" here he does not mean that the electron can be infinitely divided but that, like the atom, it can produce an infinite variety of phenomena. The infinity of the universe, he argues, in his final sentence, is not an abstract or mystical concept but an objective reality – an infinity of matter.

The "overwhelming majority of scientists," Lenin contends, have an "instinctive, unwitting, unformed, philosophically unconscious conviction" of "the objective reality of the external world." This "instinctive" materialism is being driven by the advance of science toward dialectical materialism:

> This step is being made, and will be made, by modern physics: but it is advancing towards the only true method and the only true philosophy of natural science not directly, but by zigzags, not consciously, but instinctively, not clearly perceiving its "final goal," but drawing closer to it gropingly, unsteadily, and sometimes even with its back turned to it. Modern physics is in travail; it is giving birth to dialectical materialism.[20]

Physicists will be forced to think more and more in dialectical materialist ways if they are to make meaningful generalizations. This has to some degree taken place but, as Lenin perceived, in a "zigzag" path; and in a fragmented way in the midst of idealist obscurantism. The propaganda establishment of the modern monopoly-capitalist state has so far proved too strong to permit any major trend in a consistent materialist direction. Even when physicists recognize that particles have anti-particles they do not connect this with a general conflictive quality in nature. When they are confronted with endless examples of the new (qualitative change) arising from quantitative and positional changes in the old, they fail to generalize. So, too, in the other sciences. On the other hand some prominent scientists, including J.B.S. Haldane, J.D. Bernal, Joseph Needham and Marcel Prenant, have espoused dialectical materialism, and others, like Stephen Jay Gould, have adopted some of its views. (Although Gould, for instance, recognizes change of quantity to quality and "punctuated equilibria," he fails to tie these up with the basic conflictive nature of matter.) Lenin was right in seeing that the future developments of physics would increasingly reveal the dialectical nature of matter and that this could not be properly encompassed by traditional philosophies. In the socialist world – and Lenin does not here (in 1908) seem to have had a future socialist society in mind – many scientists embraced dialectical materialism and some demonstrated how its views both aid in their researches and give general meaning to them. In the capitalist world the way is blocked by capitalist control of culture and ideology.

"The 'Struggle' of Opposites"

Materialism and Empirio-Criticism was not intended to be a general philosophical survey but primarily a study of the materialist theory of knowledge, which was then under attack by neo-Kantian idealists. In it Lenin concentrated on materialism and had little to say of dialectics. In July 1914, however, on the eve of World War I, he began an article on Karl Marx for Granat's *Encyclopedia Dictionary,* a Moscow publication. A section headed "Dialectics," following one on "Philosophical Materialism," reflects the fact that he turned (apparently in September) to a systematic study of Hegel's works (a "fog of extremely abstruse exposition," he lamented). On January 4, 1915, he wrote to the secretary of the publishers of Granat's: "I have been studying this question of dialectics for the last six weeks and I think I could add something to it [the Marx article] if there were time." In 1922, in his essay, "On the Significance of Militant Materialism," he urged Soviet scholars to "organize a systematic study of Hegel's dialectics from a materialist point of view." And this he was himself attempting in his 1914-1915 studies. His work led to a short essay, "On the Question of Dialectics," and a series of marginalia jottings on Hegel's writings, jottings now published in the *Philosophical Notebooks.* Both were written mainly for self-clarification and were not intended for publication (as his use of parens indicates).[21]

Lenin began the essay as follows:

> The splitting of a single whole and the cognition of its contradictory parts (see the quotation from Philo on Heraclitus at the beginning of Section III, "On Cognition," in Lassalle's book on Heraclitus) is the *essence* (one of the "essentials," one of the principal, if not the principal, characteristics or features) of dialectics.

Heraclitus had argued that all opposites must, paradoxically, co-exist in a unity: "For the One is that which consists of two opposites, so that when cut in two the opposites are revealed." Lenin continues:

> The correctness of this aspect of the content of dialectics must be tested by the history of science. This aspect of dialectics (e.g., in Plekhanov) usually receives inadequate attention: the identity of opposites is taken as the sum-total of *examples* ("for example, a seed," "for example, primitive communism." The same is true of Engels. But it is "in the interests of popularisation . . . ") and not as a *law of cognition* (*and* as a law of the objective world).

Plekhanov and Engels gave "inadequate attention" to the interpenetration of opposites and treated the subject too much by an "examples" method (for popular understanding). (The "seed" and "primitive communism" examples are from Engels.) Why was this wrong? The answer must be because Lenin felt that it obscures the fact that such interpenetration constitutes the essence of matter and society, and is not confined to certain specific phenomena. (Engels, as I have noted, did at times pay "inadequate attention" to contradictions, and thus inadvertently opened the gates to Stalin's idealist "rise and fall" theory, which was later accepted by Soviet philosophers in general.)[22]

Lenin's working definition of dialectics we might compress as follows: "The splitting of a single whole and the cognition of its contradictory parts is the essence of dialectics." With this we are back with Marx: "What constitutes dialectical increment is the existence of two contradictory sides, their conflict and fusion into a new category." Like Marx, Lenin's constant sense of struggle focused his philosophy on the clash of opposites.[23]

In his definition Lenin, like Marx, is not primarily discussing the dialectical process in nature but in thought – as is indicated by "cognition." The dialectical process in nature, however, does not arise, as Heraclitus believed, because "wholes" had been split but, ultimately, because matter consists of oppositional particles. Lenin was as aware of this as science at the time permitted. In the "On Dialectics" essay (1915) he lists the "opposites" in "physics" as "positive and negative electricity." True, in dialectical reasoning the basic process is indeed the discovery of "contradictory parts" but the concept of divided "wholes" in nature should be abandoned.

Lenin pursued the question of "opposites" further in his notations on Hegel:

> The identity of opposites (it would be more correct, perhaps, to say their 'unity'—although the difference between the terms identity and unity is not particularly important here. In a certain sense both are correct) is the recognition (discovery) of the contradictory, *mutually exclusive,* opposite tendencies in *all* phenomena and processes of nature (*including* mind and society). The condition for the knowledge of all processes of the world in their *"self-movement,"* in their spontaneous development, in their real life, is the knowledge of them as a unity of opposites. Development is the "struggle" of opposites.

(By placing "struggle" in quotes Lenin indicates that he is using the term metaphorically. In nature there is no actual "struggle" but a blind interaction of opposite entities.)[24]

Hegel, as Lenin noted, rejected the view (suggested in Aristotle) that all things have an absolute identity, none of which they share with anything else. If so, Hegel argued, they could not blend with other things – as they obviously did. In fact, they interpenetrated with opposites. This interpenetration Hegel sometimes called the "unity" of opposites and sometimes the "identity" of opposites – a joint identity based on their mutual dependence. As one example of this fluid identity of opposites, Lenin gave the individual and the universal: "From a certain point of view, under certain conditions, the universal is the individual, the individual is the universal." (Thus every animal possesses both the general character of its species and its own individual qualities.) Lenin prefers "unity" to "identity," and this certainly seems the more accurate term. It is confusing to say that such opposites as positive and negative electric charges are identical. Each, in fact, has its own separate identity and this identity persists alongside the mutual – and often temporary – identity resulting from interaction. However, they are certainly united – a "unity" – in their interdependence (as with the sexes).

In a particularly significant passage in "On Dialectics," Lenin views "opposites" in the context of "development":

> Development is the "struggle" of opposites. The two basic (or two possible? or two historically observable?) conceptions of development

(evolution) are: development as decrease and increase, as repetition, *and* development as a unity of opposites (the division of a unity into mutually exclusive opposites and their reciprocal relation).

In the first conception of motion, *self*-movement, its *driving* force, its source, its motive, remains in the shade (or this source is made *external*—God, subject, etc.). In the second conception the chief attention is directed precisely to knowledge of the *source* of *"self"*—movement.

The first conception is lifeless, pale and dry. The second is living. The second *alone* furnishes the key to the "self-movement" of everything existing; it alone furnishes the key to the "leaps," to the "break in continuity," to the "transformation into the opposite," to the destruction of the old and the emergence of the new.

To obscure the "struggle" of opposites" (as Plekhanov and Engels sometimes did), is to imply that development is essentially a matter of "decrease and increase." If so, we must either assume that the motive force is unknowable – and the skeptics would happily agree – or that it is basically "external," which would imply divine force (God or a Prime Mover). The "key" to "the destruction of the old and the emergence of the new" alike lies in the clash of opposites. Everything arises from this – and from nothing else.[25]

Lenin: Insight and Error

After having read many pages of Hegel, Lenin jotted down what he considered to be some of the "elements of dialectics." Among them he included: "not only the unity of opposites, but the *transitions* of *every* determination, quality, feature, side, property into *every* other (into its opposite?)."[26]

As we have seen, Engels used the concept of interchanging opposites to attack the rigid categories of formal logic: "he conception of universal action and interaction, in which causes and effects are constantly changing places, and what is now or here an effect becomes there or then a cause, and vice versa." Engels, however, did not mean that specific concepts or the phenomena they represented were actually being transformed one into the other. On the contrary, when "opposites" interacted they were transformed into something different from either component. Oxygen and hydrogen in uniting are not transformed into their "opposites" but into a third substance, namely water. And so on, as he noted, throughout the periodic table of the elements. So, too, with social phenomena. The contradiction between productive forces and productive relationship does not produce a new opposite to either but a new form of society. Lenin, however, does argue elsewhere, in spite of his doubts here – expressed by his question mark – that phenomena, social and natural, can actually be transformed into their opposites:

> That all dividing lines, both in nature and society, are conventional and dynamic, and that every phenomenon might, under certain conditions, be transformed into its opposite, is, of course, a basic proposition of Marxist dialectics. A national war might be transformed into an imperialist war and vice versa. Here is an example: the wars of the Great French Revolution began as national wars and indeed were such. They were revolutionary wars—the defence of the great revolution against a coalition of counter-revolutionary monarchies. But when Napoleon founded the French Empire and subjugated a number of big, viable and long-established national European states, these national wars of the French became imperialist wars and in turn led to wars of national liberation against Napoleonic imperialism.

However:

> Transformation of the present imperialist war of 1914–16 into a national war is highly improbable, for the class that represents progressive development is the proletariat which is objectively striving to transform it into a civil war against the bourgeoisie.[27]

The "imperialist war" of Napoleon, however, was not the "opposite" of the previous "national war" but simply a different kind of war. Is "civil war" also an opposite of imperialist war and perhaps of "national war?" If national war and imperialist war were really opposites, they would be parts of a contradictory unit, each dependent on and changing under the influence of the other. But national war is changed into imperialist war not by any such interpenetration but by the actual opposites inherent in capitalist society, namely the socio-economic contradictions that created them both. Finally, the notion of phenomena becoming their opposites leads not to developmental but to cyclical and even idealist thinking – as may be seen in some of Mao Tse-tung's philosophical absurdities: "war and peace transform themselves into each other." The whole concept of interchanging opposites is false and should be dropped.[28]

Apparently Lenin, in his eagerness to exemplify what he believed to be "a basic proposition of Marxist dialectics," simply did not think the matter through properly, something quite untypical of him. That even Lenin could lapse in this instance brings up a special problem, one that today requires emphasis, namely using dialectical "laws" as overall determinants into which facts are forced.

In his Granat's *Encyclopedic Dictionary* article on Marx, Lenin attempted a kind of summary-definition of dialectical process:

> A development that repeats, as it were, stages that have already been passed, but repeats them in a different way, on a higher basis ("the negation of negation"), a development, so to speak, that proceeds in spirals, not in a straight line; a development by leaps, catastrophes, and revolutions: "breaks in continuity"; the transformation of quantity into quality; inner impulses towards development, imparted by the contradiction and conflict of the various forces and tendencies acting on a given body, or within a given phenomenon, or within a given society; the interdependence and the closest and indissoluble connection between *all* aspects of any phenomenon (history constantly revealing ever new aspects), a connection that provides a uniform, and universal process of motion, one that follows definite laws.

Although Lenin is here basing himself on Engels and Plekhanov, he places his own emphasis and provides his own succinct vision: an infinitely interconnected reality – including both nature and society – endlessly in motion generated by "contradiction and conflict," the whole embodying a quantity-to-quality "development" in which earlier stages are "repeated" on a higher basis. Behind the total "process" lie "laws" (or, according to another translation, "law"), not subjective but objective (as other passages make clear).[29]

Lenin, here writing for a general encyclopedia audience, seems especially intent on avoiding any dogmatic-sounding presentation of dialectics and on presenting a picture of scientifically based generalization. He introduced the passage with a comparison between the Darwinian and Marxist views on "the idea of development, of evolution." Dialectics, he considered "the most comprehensive and profound doctrine of development," thus appealing (in 1915) to Russian liberal intellectuals acquainted with the "advanced" (anti-religious) Darwinian theory of evolution. Dialectical materialism was more comprehensive because it included society as well as nature, more profound because it saw contradiction as the spur to development and perceived that development assumed a quantity-to-quality form and often took place in "leaps." In his introductory emphasis on "development," however, Lenin inadvertently leaves the impression that development as such is the basic, inherent force in both nature and society.

For this and other reasons it is perhaps worthwhile to elaborate on Lenin's exposition in the light of more recent knowledge.

As I have indicated, what we call development in nature is mostly a tendency to complexity arising from conflictive units. Conflict itself arises ultimately from the antithetical nature of the elementary components of matter, as seen in negatively and positively charged particles. These opposites attract each other and form atomic and other combinations, which then form combinations of combinations. On our planet the tendency continues as atoms combine to form not only molecules but the long chain molecules of living matter. The tendency is then transferred to the animals and plants produced by the chain molecules, whose production of more complex forms is molded by the struggle for survival interacting with ecological change. But although increasing complexity is inherent in the chance encounters of opposite entities on different levels of matter and of living matter the process can also act in anti-developmental ways – as in the formation of neutron stars or the evolution of overlarge antlers. There is, then, no developmental urge inherent in matter, only blind conflict that can produce complexity. In living matter and life-forms, unless complexity results in adaptability, it is not developmental.

What then of society? The history of human – and pre-human – society shows that although both developed, there is no special developmental force at work. Societies arise and perish; some societies never develop. There are still hunter-gathering societies. However, society differs from biological nature in two major respects: its basic operating units are no longer forms of matter or living matter but socio-economic structures, and the social process has a conscious component. However, in society as in nature, for reasons I shall discuss later, the basic units are conflictive. The conflicts produced a succession of societies in various places at various times: hunting-gathering; farming; feudal; commercial capitalism; industrial capitalism; socialism.

Human hunting-gathering societies emerged from animal hunting-gathering societies, most directly from that of the early hominids (essentially upright apes). Humans, with larger brains, moved these societies forward by adding a farming component and inventing tools and weapons. When in some areas the farming component became the main one, this may have been sparked by climatic change but, if so, it was developed by societal forces. When we arrive at farming society we reach one of those points in a sequence of change in which certain other changes become inevitable or, as Engels would say, necessity emerges out of chance. Farming by its nature produced a much greater surplus of goods than even the richest hunting-gathering societies had and hence, developed trade on a new level (as for instance, the trade that must have created the walled city of Jericho some 8,000 years ago). The larger farm holdings necessarily became still larger in the competitive struggle and the smaller still smaller, until we have exploitive feudal lords at one end and the serfs that worked their land at the other. In short, people became divided into classes, depending basically on their position in the economy, and classes, reflecting conflictive economic and other interests, joined in class struggle.

Feudalism, from Egypt to China, developed much greater agricultural surplus than even the most advanced farming societies and hence much greater trade. Then, as commercial wealth – especially when traded by water – grows much more rapidly than agricultural wealth it was but a matter of time, historically speaking, before the commercial component of feudal society became dominant. Then after many ups and downs it emerged as a new society in its own right, that of commercial capitalism, which, largely for geographical reasons, emerged first in Europe.

Because of cosmic accidents – supernova explosions – several billion years ago that shot iron and other metals into our planetary rock and then the geological mashing of forests, beginning a mere 300 million years ago into coal, oil and natural gas, commercial capitalism was able to turn into industrial capitalism.

In industrial capitalism as in commercial capitalism and feudalism, the big economic units became bigger until industrial capitalism assumed a corporate (monopoly) form. In this form it branched outward with new impetus until it had colonized virtually all the rest of the world, a situation which produced inevitable conflicts between the corporate capitalist powers. These conflicts, then, accentuated both the class struggle within the system and that between it and the colonies and semi-colonies. This situation produced social explosions that led to socialism in various parts of the world, a socialism now fighting for its life against imperialist pressures.

Thus the emergence of industrial capitalism, although dependent on natural accident, marked another turning point from which a certain line of development became inevitable.

Thus although there is not, as some Marxist and Soviet texts have claimed, a drive to development inherent in human society in general, there are, as I have noted, certain points at which events can move in only one direction, as in an embryonic development. But whereas in nature these things work their way out through mechanical trial and error, in society they are carried out by a series of planned actions of a limited nature: building factories, establishing parliaments, waging peasant revolts. These actions resulted in non-planned major change. In fact until industrial capitalism created a working mass operating a concentrated economic system, major historical change could not take place by overall conscious human direction. The first time this happened was in the Russian revolutions in 1917. Nobody sat down and planned the transition from feudalism to commercial capitalism or from commercial to industrial capitalism. But the transition to socialism, once it was made possible by the conditions of the age, was consciously planned. Humans, at last, had control of basic social forces. The direct instruments for this control were Marxist-Leninist parties which, as such, had a scientific understanding of the workings of society. Human action, directed by such parties, can prevent nuclear war and ecological devastation and clear the path for the continuing rise of socialism.

In spite of the differences between nature and society, societal development takes place in the same general way as in evolution, that is to say, out of a " 'struggle' of opposites."

Like Engels, Lenin does not indicate why nature and society should act in the same general way, namely by qualitative change arising from quantitative-arrangement change in response to the clash of opposite entities. Without at least a brief explanation – namely that people emerged from nature and human society from animal society – the parallels seem almost miraculous.

As also with Engels, we have in Lenin the "negation of the negation" producing development on a "higher basis." Finally, again as with Engels, the concepts of "motion" as a basic force and of objective "law." It is clear today that motion is not basic but derivative. Nor, as I have argued, are there "laws" at work in the universe or in history, but only processes. Lenin's final "uniform, and universal process . . . that follows definite laws" inadvertently has idealist implications and gives an almost Newtonian sense of orderliness under celestial law – a very different picture from that of the "violent universe" of blindly clashing forces revealed by physics and astrophysics today (but hardly being hinted at when Lenin wrote).

Some errors made by Engels were taken over and magnified by Stalin, and then entered into Soviet philosophy. Lenin, however, has to bear some of the blame for some of this, not so much by anything that he himself expounded but by occasional loose formulations and a failure to note divergencies between Engels and Marx. For instance, in his *Granat* encyclopedia article on Marx he quotes the rise and fall passage from *Feuerbach:* "the world . . . is a complex of processes, arising here and disappearing there [resulting in an] unending ascent from the lower to the higher." He fails to indicate that the passage – essentially, as Engels notes, an exposition of Hegel – leaves out the motivating essence of contradiction and could imply teleological progress in society and nature. After quoting the passage, Lenin comments: "Thus dialectics, according to Marx, is 'the science of the general law of motion both of the external world and of human thinking.'" This quote, however, is from Engels and not Marx, and the implication – "thus" – is that Marx would have agreed not only with it but with the just quoted long "rise and fall" passage. The next sentence runs: "This revolutionary side of Hegel's philosophy was adopted and developed by Marx. Dialectical materialism 'does not need any philosophy towering above the other sciences.'" This quote also is from Engels, not Marx. The impression given, then, is that there are no differences between Marx and Engels. The one speaks for the other. It is also implied that Marx believed that Hegel's exposition of dialectics had a "revolutionary side." But Marx denounced Hegel as a reactionary proponent of dialectics in its "mystified form" (applauded by the Prussian establishment) and there is no indication that he regarded Hegel's dialectical "method" as in any way revolutionary. Lenin here by implication echoes Engels' rather approving view of Hegel rather than Marx's basically critical one. So too, in the opening sentence of the *Dialectics* section: "Marx and Engels regarded Hegel's dialectics, the theory of development most comprehensive, rich in content and profound, as the greatest achievement of German classical philosophy." Hegel, however, as Marx recognized, was not propounding a scientific developmental theory but dialectical mysticism. Similarly by the exclusive attention he gave to Hegel in the *Philosophical Notebooks* – not intended for publication – Lenin inadvertently gave ammunition to those who wished to increase Hegel's stature and give an exaggerated impression of his influence on dialectical materialism. Lenin makes no reference to the important dialectical component in such materialists as Lucretius and Diderot. Perhaps he was unaware of it.[30]

In Appendix II of my *Stalin: Man of Contradiction* I expressed puzzlement over Lenin's formulations in two passages which Stalin paraphrased with something of an idealist spin. The first occurs in Lenin's *Granat* article on Marx (1914), the other in a short essay, "The Three Sources and Three Component Parts of Marxism" (1913):

> If materialism in general explains consciousness as the outcome of existence, and not conversely, then materialism as applied to the social life of mankind must explain social consciousness as the outcome of social existence.

> Just as the cognition of man reflects nature (i.e., developing matter) which exists independently of him, so also the social cognition of man

(i.e., the various views and doctrines, philosophic, religious, political, etc.) reflects the economic order of society.

Marx, of course, means that there is only one consciousness (or "cognition"). This consciousness, although formed by social forces, reflects nature as well as society. And this is certainly what Lenin means also but some of his formulations, especially in the 1913 passage, seem murky. In it he seems inadvertently to confuse cognition (consciousness) as such with its contents. There is no "cognition" as such and also a "social cognition" as such. The two are one and the same. Only their contents are different.[31]

The Curve of Knowledge

Although Lenin does not specifically examine Marx's *ad Feuerbach* notes as providing a new base for epistemology (knowing) in a fluid, interacting process of thought, sensation, emotion and practice, he develops a similar view. In *Materialism and Empirio-Criticism,* as we saw, he emphasized the relationships between "external excitation," "sensation" and "consciousness." In the *Philosophical Notebooks,* he takes the matter further:

> The *reflection* of nature in man's thought must be understood not "lifelessly," not "abstractly," *not devoid of movement,* not without contradictions, but in the eternal process of movement, the arising of contradictions and their solution.

Although Marx had stressed the interaction of the elements involved in the cognitive process, he did not specifically note, as Lenin does here, its inherently dialectical nature. This involves, to quote Lenin, a "zig-zag like" motion of opposites including sensation as well as thought: "not only is the transition from matter to consciousness dialectical, but also that from sensation to thought." He makes the basic point that thought is itself not a form of matter but is rooted in matter: "concepts are the highest product of matter." In another passage Lenin goes a step further: "Man's consciousness not only reflects the objective world, but creates it" – "creates," of course, not absolutely but imaginatively.[32]

Both the simpler mechanisms of "sensation" and the more complex ones of conscious thinking, Lenin saw as arising from an intermingling of opposites. Nothing was then known of the specific processes at work but some beginnings had been made. Wilhelm Max Wundt had established the first psychology laboratory in Leipzig in 1878. "Sensations," such as pain, were being examined experimentally. William James and others argued that the emotions had a physiological basis and considered consciousness to be "active." Such developments doubtless provided material for Lenin's views. But the views themselves with their central emphasis on the contradictory essence of psychological process represent a new level of materialist understanding. Today when the basic mechanism of the brain has been shown to embody a series of interacting opposites, including negatively and positively charged atoms and the excitative and repressive molecules arising from them, it becomes apparent that Lenin's general grasp of contradiction again led him in the right direction.

By the comment "concepts are the highest product of matter," Lenin means, as he makes clear elsewhere, "organic" matter, the matter of life. To fail to note this in expositions of the subject, as some Soviet philosophers did, is, in effect, to leave out evolution and give the impression of matter somehow organizing itself into a "higher" form or even having consciousness somehow inherent in it. Today we are beginning both to uncover some of the specifics of the bio-psychological process involved and to perceive the long evolutionary trail – from amoeba to mammals, from simple reflex to behavioral patterns – behind it.[33]

Although Marx ridiculed the false abstractionism of Proudhon and all who tried to fit reality into "categories," he felt that abstract generalizations based on concrete reality (as he demonstrated on a large scale in *Capital*), provided a major path to truth. Lenin thought similarly:

> The abstraction of matter, of a law of nature, the abstraction of value, etc., in short all scientific (correct, serious, not absurd) abstractions reflect nature more deeply, truly and completely. From living perception to abstract thought, and from this to practice, — such is the dialectical path of cognition of truth, of the cognition of objective reality. Kant disparages knowledge in order to make way for faith: Hegel exalts knowledge, asserting that knowledge is knowledge of God. The materialist exalts knowledge of matter, of nature, consigning God, and the philosophical rabble that defends God, to the rubbish heap.

Abstractions derived by materialist reasoning from the concrete – "living perception" – to the general is the form of thinking that brings us closest to the essence of nature (or society). (And Bacon would have agreed.) The second movement, from abstraction down to "practice," cannot, of course, stand alone. Nor did Lenin intend it to. As he noted in another place: " . . . man by his *practice* proves the objective correctness of his ideas . . . " Here, however, he notes the reverse of this process, namely that abstractions formed by practice assist practice – a necessary (dialectical) addition to Marx's and Engels' statements on "practice" as the basic criterion for truth, an addition they would certainly have accepted.[34]

Science is not only a matter of experiment or logical deduction: "it would be stupid to deny the role of fantasy, even in the strictest science." Einstein, we might note, said that he developed his theory of relativity in part by pretending to be a light particle travelling through space and imagining the forces that would act upon him. And other scientists have made similar statements. But it is interesting to find Lenin, who is often represented as an incarnation of revolutionary logic, acknowledging the value of fantasy. Lenin, as is too seldom recognized, was not only a disciplined revolutionary but a man of deep feelings and imagination. The general question he raises of fantasy as a path to truth, however, goes beyond science, natural or social. In fact it is not mainly connected with science but with the arts. Artistic creation, as I have suggested, has its own paths to the truth; but these must be based, as Lenin indicates for thought in general, on "living perception" and "practice" unimpeded by religious delusions. Let us take, for instance, Shelley's fluid vision of the universe, which compresses creatively – and so exceeds – that of the astronomy of his day:

> *Worlds on worlds are rolling ever*
> *From creation to decay,*
> *Like the bubbles on a river*
> *Sparkling, bursting, borne away.*[35]

Philosophers have long speculated about the difference between "appearance and reality," sometimes with the implication that "appearance" is a delusion and "reality" is either the manifestation of God or known only to God (of which Kant's ever unknowable "thing in itself" was a sophisticated version). With the rise of science, the controversy acquired new meaning, for science revealed a very different aspect of things than the senses did. This was recognized by Marx: "Scientific truth is always paradox if judged by everyday experience, which catches only the delusive appearance of things." With the discovery of the electron and radioactivity, the dichotomy became even more acute and idealists used it to support the argument that not only matter but all "appearance" was a delusion. Lenin answered as follows: "The essence here is that both the world of appearances and the world in itself are moments of [steps in] man's knowledge of nature, stages, alterations or deepenings (of knowledge)." Or, as he put it metaphorically: "the movement of a river – the foam above and the deep currents below. But even the foam is an expression of essence." The fact that the world as revealed by science, the world of electrons and atoms, was very different from that revealed by everyday experience did not mean that the first was real and the second unreal. Both are aspects of the same reality – the "foam" as well as the "deep currents." The metaphor, we might note, shows Lenin's talent for imaginative expression, which like his comment on fantasy reveals his feelings for artistic creativity (as seen, for example, in his writings on Tolstoy).[36]

The world as it appears to the senses, then, much though it may differ from that revealed by science, is not – as Marx suggested – "delusive" and the world of science "real." To put the matter simply, as my wife Mary Bess Cameron did when I discussed it with her: "If we look at this tree and then examine parts of it through an electron microscope, we are still seeing a tree."

Lenin takes the problem still further: "it is equally incorrect to regard the Idea as something "unreal" – as people say: 'it is merely an idea.' " Because an idea is subjective and not objective does not mean that it is any the less real. Even "fantasy" is real. Dreams are real. Both subjective and objective are forms of reality; thoughts and emotions, as much as atoms or social phenomena. Lenin is unearthing an idealist attitude. If dreams are "unreal," what is their source?[37]

Lenin's general position holds still further implications. As we have seen, Lucretius was able to anticipate, even though in a general and fragmented way, some of the findings of modern science. He did so in part because of his knowledge of the meager science of his day but mostly it came from simple observation. Lenin might have said that Lucretius was on the right track because he reflected aspects of the same reality as that later unveiled by the instruments of science. The situation was similar with Hegel and dialectics. By Hegel's time the American Revolutionary War and then the French Revolution had revealed a developmental stream in history. Kant had projected a developing universe. Chemical and electrical experiments had begun to indicate that matter acted, as the Greek materialists had argued, by the resolution of "opposites." These experiments, of course, only skimmed the surface, but the surface reflected some aspects of the deeps. Hegel grasped the general (dialectical) vision evoked by these advances and turned it into an idealist credo.

Hegel, like Kant, saw reality in terms of categories. Lenin noted his views, sometimes questioning them, sometimes developing them. In humanity's long search for understanding, the creation of such "categories" as cause and effect or individual and universal marked a step forward:

> Man is confronted with a *web* of natural phenomena. Instinctive man, the savage, does not distinguish himself from nature. Conscious man does distinguish, categories are stages of distinguishing, i.e., of cognising the world, focal points in the web, which assist in cognising and mastering it.[38]

Lenin, here partly paraphrasing Hegel, adds with approval Hegel's comment that categories though useful can be "instruments of error and sophistry." "The categories," he continues, speaking of Hegel's categories, "have to be derived (and not taken arbitrarily or mechanically)." They must blend into one another because the natural phenomena they reflect are interactive: "If everything develops, does not that apply also to the most general concepts and categories of thought? If not, it means that thinking is not connected with being. If it does, it means that there is a dialectics of concepts and a dialectics of cognition which has objective significance."[39]

Marx, as we saw, did not regard categories as differing in any major respect from "ideas" or "principles." So, too, Lenin. Categories are simply abstractions, "general concepts," not unique forms of thought. The early "savage" blurred subjective and objective. As "man" achieved consciousness, he began to disentangle the "web" by arranging its parts into "categories," "general concepts," which like all concepts are valid only if they reflect reality, are "connected with being." Like "being" itself they develop. They cannot, as Marx said of Proudhon's economic categories, be "eternal" or frozen or imposed on objective reality. Whether the abstractions of dialectical thinking with its fluid reactions – of cause into effect and effect into cause, for instance – can be considered categories in any meaningful sense, seems, as I argued in relation to Engels, doubtful. Lenin does not raise the question. On the other hand, not only does he not consider categories as a special logical entity but he gives no sign of attempting to "master" or even to form a "system" of them. It seems doubtful that he would have raised the question at all if he had not been reading and commenting on Hegel, for there is virtually nothing about categories in his other works. Unfortunately simply by paying as much attention as he did to Hegel's views on categories, he inadvertently played into the hands of later Soviet philosophers who made a fetish of them, turning dialectical materialism into dialectical idealism.

In his description (quoted above) of dialectical process in his

Granat encyclopedia article on Marx, Lenin included its revelation of an "indissoluble connection between *all* sides of any phenomenon (history constantly revealing ever new aspects)." In the *Philosophical Notebooks* he pursues the concept:

> Human knowledge is not (or does not follow) a straight line, but a curve, which endlessly approximates a series of circles, a spiral. Any fragment, segment, section of this curve can be transformed (transformed one-sidedly) into an independent, complete straight line, which then (if one does not see the wood for the trees) leads into the quagmire, into clerical obscurantism (where it is *anchored* by the class interests of the ruling class).[40]

Human knowledge assumes a spiral form because society has developed in a series of spirals, with ups and downs within a central upward curve. Non-dialectical thinkers can take segments of this spiral and artificially transform them into straight lines with a beginning and an end, thus cutting them off from their interconnections. By excluding some of the findings of science they can come up with a "clerical obscurantism" that helps a "ruling class" to keep its working mass subdued. In fact, Lenin suggests, one of the characteristics of bourgeois (or other exploitive-class) thinking is to reason within narrow areas because historical interconnections, if properly explored, reveal the exploitation and oppression at the heart of the society. In contrast materialist-dialectical thinkers dig to reveal the very connections that bourgeois thinkers obscure, including natural (philosophical) as well as social ones.

In a debate with Nicholai Bukharin in 1921 on the role of trade unions in a socialist state, Lenin pursued the subject further:

> Formal logic, which schools confine themselves to (and which, with modifications, the lower forms should confine themselves to), takes formal definitions, and is guided exclusively by what is most customary, or most often noted ... Dialectical logic demands that we can go further. In the first place, in order really to know an object we must embrace, study, all its sides, all connections and "mediations." We shall never achieve this completely, but the demand for all-sidedness is a safeguard against mistakes and rigidity. Secondly, dialectical logic demands that we take an object in its development, its "self-movement" (as Hegel sometimes puts it), in its changes ... Thirdly, the whole of human experience should enter the full "definition" of an object as a criterion of the truth and as a practical index of the object's connection with what man requires. Fourthly, dialectical logic teaches that "there

is no abstract truth, truth is always concrete," as the late Plekhanov was fond of saying after Hegel . . . [41]

Lenin, like Marx in his discussion of his "method," places the emphasis first not on dialectical logic but on factual examination and establishing interconnections. But he goes somewhat beyond Marx in his particular emphasis on the necessity of taking an all-sided view – "study all its sides" – in examining natural or social phenomena. These, of course, are not static but in motion. Hence we must "take an object in its development," seeing not only interconnections but roots and movement and, if possible, future course. We might take Lenin's own *Imperialism: The Highest Stage of Capitalism,* as an example. He there examines monopoly capitalism in its interconnections, economic, social and political, both within itself and in colonial or other imperialist relations. He views it as a developing system, arising out of industrial capitalism. He reveals its movement, as a consequence of internal oppositional dynamics, towards greater monopoly and imperialist war.

Lenin, of course, does not mean that we should take the whole of human experience into account in attempting to get to the "truth" of any process but only that we should use what of it is relevant as a general foundation. The final comment on truth always being concrete and not abstract may at first seem puzzling after Lenin's comments in the *Notebooks* on the value of abstract thinking. What he must mean is that behind abstraction there must lie concrete facts if it is to be valid.

Although "causation" is based on objective phenomena, the projection, as with all such projections, is imperfect:

> The all-sidedness and all-embracing character of the interconnection of the world, which is only one-sidedly, fragmentarily and incompletely expressed by causality.

Lenin here develops what is perhaps inherent in Engels' argument that "cause and effect" are "conceptions" that reflect a "universal action and reaction" that "run into each other." Lenin notes (typically) that this reflection is necessarily one-sided, fragmentary and incomplete. Is causality, then, inherent in nature? Lenin implies, perhaps more strongly than Engels, that it is not. What we have in nature is "interconnection." The interconnections of phenomena are translated by the mind into causality: "expressed by causality."[42]

Although Lenin did not, anymore than did Marx and Engels, put together in one place what he meant by materialist-dialectical thinking, it is not difficult to do so. Like Marx he began not with dialectics as such but with the materialist base, repudiating at the outset the new subjective idealists and skeptics. In doing so he relied primarily on "practice," including that of the new physics. The essence of matter was "the 'struggle' of opposites" – as exemplified in negative and positive electricity. Because of this "struggle" (in quotes), matter is developmental. So, too, is society. Hence, we have to think about both in terms of "contradictions" and the interactions following from them, seeking out all their "aspects," tracking down – as Marx also stressed – the "connections" of any phenomenon, natural or social: "All-sidedness is a safeguard against mistakes and rigidity." Practice provides verifications; not absolute verification but close enough to sustain further practice, either in science or political action. In these proceedings reason blends with imagination. Even "fantasy" can help clear a path to the truth. If we think in these ways, we think like dialectical materialists. Lenin did not, of course, arrive at these views by "pure reason." Like Marx he was a revolutionary before he was a dialectical materialist and he was guided by that omnipresent sense of "struggle" that gave him depth of insight.

Lenin expanded the views of Marx and Engels in accordance with the outlook and methods he had learned from them, much as they – especially Marx – would have done had they lived in his world. In reading his philosophical works, we are immediately struck by their combination of creative insight and systematic investigation. He was an extremely responsible thinker, assessing his materials thoroughly and honestly, hunting only for the truth and without egocentric motive in the quest. He went through the "abstruse" "fog" of Hegel's *The Science of Logic* and *Lectures on the History of Philosophy* line by line, underlining passages and making numerous marginal notes and queries. When he was not clear on Hegel's meaning, he did not hesitate to say so. When he was unsure of his conclusions he put them tentatively or followed them by a question mark. None of this as a mere academic study but in order to improve his use of dialectical reasoning in his elucidation of everything from imperialism to everyday struggles.

Although he erred on some matters, in general he raised dialectical materialism to a new level of sophistication. In *Materialism and Empirio-Criticism,* he expanded its materialist base, indicating that the discovery of the electron and radiation did not mean that "matter has disappeared" but only that it had a greater richness of forms than had been thought. At the same time he demonstrated that idealism and skepticism, far from being intellectually acceptable, could alike be reduced to solipsistic absurdity and, moreover, were inherently reactionary whether their proponents realized it or not. In *On Dialectics* and the *Philosophical Notebooks* he reestablished the central role of conflictive forces in nature and society, demonstrating as he did so the reactionary implications of the theory of "development" by "decrease and increase." He raised dialectical-materialist epistemology to a new level, as in his "man in a dark room" analogy or his theory of the dialectical "zig-zag" processes of "knowing" – between matter and the senses and between the senses and emotions and thought – a view which is being validated by scientific research today. He demonstrated further the "metaphysical" (and false) character of the view of mind as passively receptive. He brought out what was only implied in Marx and Engels, namely that thought is not itself material but a "product of matter"; and this, too, is being demonstrated by modern science. He argued – again going beyond Marx and Engels – that "appearance" is not a "delusion" but an aspect of reality. He showed that abstraction reacts on "practice," thus completing the dialectical interchange. He demonstrated in new detail the difference between "formal" logic with its fragmented approach and dialectical logic with its all-encompassing interacting spirals. In *Materialism and Empirio-Criticism,* he brought materialism up-to-date in the light of scientific progress. Clearly, the same thing has to be done today.[43]

Chapter Four

MATTER, THE UNIVERSE AND LIFE*

There seems to be little attempt today to interpret the findings of science from a materialist point of view. In the capitalist world there has been no major follow-up of the efforts of such Marxist scientists as J.B.S. Haldane or J.D. Bernal to present the larger picture. The emphasis is on specialized articles. In the Soviet Union the field was long dominated, as I have indicated, by idealists. In the following chapters I attempt to do what those more knowledgeable than I in these matters should have done but have not. In these chapters I take basic works of scientific exposition from Steven Weinberg to Stephen Jay Gould, and try to integrate what we now know of matter, living matter and people into an ever-expanding dialectical-materialist view. Future developments will, of course, render much of this obsolete but, as with the 19th century science explicated by Engels, much of it will remain valid even though within new perspectives.

The Nature of Matter

As we have seen, our concept of matter has changed considerably since the days of Engels. It has, however, changed in ways that he would have found congenial; for instance, the discovery that the elementary particles have anti-particles and that the protons and electrons in the atom act in response to positive and negative electric charges. Even the presumably stable core of the atom (the nucleus) is constantly in motion, with its protons and neutrons (uncharged protons) reacting as opposites, approaching and receding in billionths of a second.

*This chapter and the next were commented upon in their various stages of evolution by my friend of more than sixty years, Dr. Rachmiel Levine, Emeritus Medical Director of the City of Hope Medical Center in Duarte, California, who has extraordinary capacity for explaining science in simple terms. He has cleared up a lot of things for me, especially in the biological sciences, and weeded out a number of errors. He still questions my interpretation on some matters, for instance on body-mind relationships (which he views in rather more strictly biological terms than I do).

Rutherford's theory that the (negatively charged) electrons swirl in orbits around the (positively charged) nucleus – a little like bees around a hive – has been borne out; and it has been discovered that the electrons move within orbital rings – spinning as they orbit – and have (relatively) immense spaces between them. The diameter of some atoms is 100,000 times that of the diameter of the nucleus. An electron can jump from one orbit to another but not remain between orbits. And it moves instantly. In short, change at the roots of atomic matter is not gradual but, as Plekhanov in particular might have delighted to note, sudden, in "quantum" leaps. And it is oppositional in its essence, for negative and positive electrical charges attract each other, but positive repels positive and negative repels negative. This simple mechanism controls not only atomic matter but living matter also, spiraling simplicity into complexity like a computer. Atoms, drawn together by their positive-negative reactions, form molecules, and groups of molecules link to make the "chain molecules" of living matter.

The viable reality is, of course, not "the atom" but *atoms,* in an electric blaze of clashing opposites as inner ring electrons are drawn still farther inward by the (positive) protons in the nucleus, and outer ring electrons are driven farther out by the negative charges of the inner ring, creating a volatile whole ready to unite with other atoms.

If we could view the elements on the screen of a supermicroscope, the immediate difference we would see between them would be that atoms of gas are much further apart than those of solids or liquids. Whereas the atoms of liquids or solids would appear almost in contact, those of gases would be far apart, yet still combining or bouncing off each other. If we witnessed a gas changing into a liquid, we would see the atoms drive closer together. And the elements which appear to our unaided eyes to be so different, for instance, oxygen and iron, are actually different only in the number and arrangements of their atomic particles and the spaces between their atoms. In short all the variety and wonder of the world of nature around us, from mountains to ocean, simply reflect differences in the number and arrangements of the elementary particles and their atoms.[1]

Matter on Earth, however, as the earlier materialists did not know and which first became apparent in early 19th century electrical experiments, is not homogenous but exists in different forms, from molecules and atoms to streaming particles (electricity). Moreover some 90 percent of the observable matter in the universe contains no atoms or molecules but exists in a less organized state known as "plasma," in which atomic nuclei endlessly collide with each other –

positive repelling positive – and electrons are a free-floating stream. In galactic gas clouds, plasma exists as a gas. In the center of our sun and other stars, it is dense, compacted by gravity.

Throughout the universe, as on Earth, matter also exists in the form of streams of elementary particles, mainly photons, the particles of light but also of electrons and neutrinos (which seem rather like uncharged electrons). Although photons are often considered a distinct entity, a unique "radiation" – which some endow with deity-like powers – they are clearly elementary particles. They not only interact with protons and electrons but can, in agitated particle states, combine to form them. In their compressed form as laser beams they act on clumped matter with considerable force (thus clearly demonstrating their material nature).

Photons, like electrons, are all the same. Their qualitatively very different effects come solely from the quantitative factor of different wave lengths. If they occur in short-wave frequencies, they form gamma rays or X-rays; in moderate frequencies, ultraviolet, visible light or infrared; in low frequencies, microwaves, radio, radar and television. All consist only of photons and all, regardless of wave-length, travel at the same speed, the speed of light. Although visible light appears to the eye to be continuous, it is, as Gerald Feinberg puts it, actually "granular," the impression of continuity coming from particles streaming in immense numbers and at high speed. Differences in what we perceive as color result from minute differences in the number and degree of compression (wave-length) of the photons of visible light.[2]

Photons can be created by the actions of electrons, in which they can live like ghostly embryos. When an electron moves from an outer to an inner ring it emits a photon. The light that drenches our world and permits us to see consists of photons made by the clashing and fusion of elementary particles in the sun hurled in trillions upon trillions into space. Conditions in other stars (suns) produce even greater particle agitation and create short wave photons such as X-rays. A lesser degree of cosmic agitation spreads the photons into long waves, such as radiowaves. Thus quantity produces quality – the new – in ways not dreamed of by Marx, Engels or Lenin.

In view of all we know about the universe, it is somewhat startling to realize that almost all of it comes from the analysis of photons mainly from stars and galaxies. There is nothing mysterious either about the universe or our means of learning about it. All is understandable and we are understanding more and more.

When physicists first began to collide particles under magnetic force in accelerators, a bewildering array of new particles appeared, leaving people with a feeling of hopelessness before the apparently unfathomable complexities of science. Most of these particles, however, exist only fleetingly, some of them the opposites (anti-particles) of the basic particles and no longer playing a significant part in the universe. In time the picture cleared further when it was perceived that particles fell into a few major groups. The electron, for instance, belongs to a group of indivisible particles called leptons; where it is joined by the neutrino, which penetrates all but the most dense matter and daily passes through the Earth – and us – in immense numbers. Similarly photons were found to belong to a group called bosons, which, unlike electrons, can "bunch up and merge." Protons, neutrons and leptons are all thought to consist of smaller particles called quarks. True, quarks have not yet been detected but skepticism pales before the exceptional track record of theoretical physics in recent decades. Many particles were first predicted and then discovered. Such theoretical speculation is, of course, not simply guesswork but is based on calculations from known phenomena.[3]

It was found also that particles interact more deeply than had been realized. In high agitation conditions – as in stars – a proton will turn into a neutron if an electron joins it (as the positive and negative charges cancel each other). Conversely, a neutron can become a proton by emitting an electron. The picture is one of incredibly fluid activity based on conflictive interactions.[4]

Although modern science reveals an extraordinary complexity of detail, then, it reveals increasingly simple underlying patterns. The three groupings of particles, for instance, can be reduced to two: those based on quarks – protons, neutrons and electrons – and those with no such base, namely photons. Linked to these particles are what some scientists call "forces" but are now, more correctly, generally called "interactions": gravity, electromagnetism, "strong" nuclear (holding protons and neutrons together in the atom's core and thus forming the base for the atomic bomb), and "weak" nuclear (seen in the decay of uranium and other radioactive elements). These are not independently creative "forces" but simply different kinds of interactions arising from different kinds of particles.

"Force," as Marcia Bantusiak notes, although often depicted as a "kind of invisible entity that pushes or pulls us around," is actually rather like a "tennis game" in which a "force between two particles arises from their continually exchanging another identifiable particle (a subatomic tennis ball, so to speak)." "For electronic interactions, 'the tennis ball' is the photon" (popping out of the electron). Similar particle interactions have been demonstrated in the weak nuclear force whose "tennis balls," the W and Z particles, were first hypothesized and then discovered. And basic particles have been hypothesized for the strong nuclear force ("gluons") and gravity ("gravitons"). The whole array of the universe, then, apparently arises from quantity-arrangement changes among the elementary particles of matter, particles whose basically oppositional character is shown by the fact all except the photon, have anti-particles; and the fluid, "bunching" photon is said by physicists to act as its own antiparticle.

This basically antithetical character of the elementary particles is not a simple set reaction but manifests itself in different ways in different situations. For instance, an electron can act as an opposite to a proton, to a positron (an anti-electron, containing a positive charge) and to a photon. The degree of interactivity between particles was further revealed in the discovery that under conditions of intense particle activity (in an accelerator) two "forces" that seem, to quote Nigel Calder, as different as a "jet plane" and a "gas cooker," namely the electromagnetic and the nuclear weak force, can blend into one. The intermediate particles that constitute this interaction were also first hypothesized and then found.[5]

Twentieth century science, then, has uncovered depths of matter unimagined in the 19th century. But they have all proved to be of a materialist and dialectical nature – of blindly clashing entities producing qualitative change – the new – by quantitative and arrangement change, and nothing else. The old argument as to whether "motion" was inherent in matter has long gone by the boards. Behind motion lie the elementary particles and their "interactions." There is, demonstrably then, no need to hypothesize a supernatural entity bestowing motion on matter.

The Nature of the Universe

In 1755 Emmanuel Kant, basing himself on the work of an English astronomer, Richard Wright, speculated that sometimes what looked like a "star" might be a galaxy like our Milky Way but "situated so far away from us that even with telescopes we cannot distinguish the stars composing it." It is only in the past few decades however, that the immensity glimpsed by Kant has begun to assume detail. There are about 100 billion galaxies, some spirals with projecting arms, some elliptical, resembling elongated saucers, all curiously flat. The galaxies are herded in clusters by gravity and the clusters into superclusters, one of which contains twenty clusters and "stretches a *billion* light years from end to end." Massive though the galactic system is, it seems to occupy only one, comparatively small, segment of space. Vast areas of apparently empty space, some of them capable of holding 5,000 galaxies, have been detected. The galactic clusters are in general moving apart from each other "like," to use Marcia Bartusiak's homely simile, "raisins in an unbounded loaf of raisin bread," and at the same time the larger clusters are pulling the smaller ones toward them. Scientists on Earth are able to detect these movements because photons from a body moving away from us are stretched out (the "red shift") but in one coming towards us are bunched up. So, too, with the movements of quasars, "quasistellar objects" in the far reaches of space that are apparently the "hyperluminous nuclei" that characterize some galaxies.[6] What we call the Universe, then, is essentially a series of galaxies which, although in different stages of growth, are basically the same, operating in the same mechanical way, as though an undirected celestial beltline had turned out trillions of copies of the same model.

The Milky Way, of which our solar system is part, is a spiral galaxy 100,000 light years across but only 2,000 light years thick – a thin pinwheel in space. It contains about 100 billion stars, some larger than our Sun, some smaller, with our solar system out in one of the spiral arms in an area populated by relatively recently formed stars. Beyond this area is a circle of ancient stars, some of which are thought to be 18 billion years old. In addition to the stars, there are clouds of dust and gas (that contain the molecules of water, ammonia and carbon monoxide). Some of the gas is in the form of large bubbles of "hot" plasma out of which stars are apparently being molded, star-making that seems to come in "periodic spurts."

When stars form, their lives become a contest between their internal actively expansive electromagnetic plasma pushing outward and the "opposite" external force of gravity pushing inward. These forces, that give them their vitality, also, in time, kill them. The nature of their demise depends on their size. Compacted by the relentless drive of gravity, large stars explode, or rather, "implode," smaller ones fade into a "dense ash." Similarly whether a body forming in space becomes a large planet, like Jupiter, or a star, like our sun, depends simply on the amount of matter it collects. Stars are also acted upon by external events; for instance, "the violent dynamics at the heart of dense starry clusters," which can strip them of their cloak of gases, as stars react to each other.[7]

Radio telescopes and other instruments reveal surges of electromagnetic agitation in the center of the galaxy. It is theorized that this might result from gravity swirling matter towards a "black hole," which, of course, is not a "hole" at all but – if it exists – a star or stars so highly compacted by gravity that their particles cannot clash to generate photons. But the discovery of a "horseshoe-shaped plume of radio emission" rising 700 light years above the flat galactic disk might, some believe, point not to a black hole but to a directly electromagnetic source: "Magnetic lines of force coiled within the plane of the disk like a cable wire wrapped around a spool" could be compressing and then expelling streams of charged particles.[8]

Electromagnetic forces may also provide the root mechanism for the formation of stars and galaxies from gaseous matter. Gravity cannot unite particles but acts only on clumped matter. How, then, does the matter become clumped in the first place? As electromagnetic action can be shown in lab experiments to clump particles, it is argued that it performs the same function in space, clumping gaseous material to a point at which gravity can take over. Others go further and argue that electromagnetic forces may account for cosmic phenomena generally attributed to gravity.[9]

As seen by the naked eye and even by early telescopes the universe appears serene and orderly – the Newtonian vision prevalent into the 1920's and often taken as representing a divine serenity. But by the 1930's the image began to change, as it became apparent that the universe was a swirl of endlessly conflicting forces. In 1936 when the physicist, Max Born, published a general cosmological study, he called it *The Restless Universe.* When Nigel Calder published a similar study in 1969 he titled it *Violent Universe.* Today the "violence" is seen to be more extreme and on a vaster scale than was apparent in 1969. In addition to "supernova" star explosions, from time to time the internal turmoil of galaxies is escalated by collisions with other galaxies. One such "cataclysmic" collision "not only dragged hordes of stars out of each galaxy but the resultant shock waves sparked a round of star formation that continues to this day." Galaxies also, for unknown reasons, sometimes explode. The universe is conflictive in its essence.[10]

The Evolution of the Universe

Although by the 1940s the theory that the galaxies were moving away from each other had been largely accepted, no reason for the expansion seemed apparent. It could hardly be gravity, it was reasoned, for gravity in the present universe does not force bodies apart but, on the contrary, pulls them together. George Gamow and other scientists early suggested that the very early matter of the universe had been highly compact but had exploded and the galaxies were still being propelled by the force of this explosion (which Fred Hoyle, who favored a "steady state" universe, derisively dubbed "the Big Bang," a name that stuck). It was also suggested that traces of the "radiation" from this primordial "bang" should still be present in a diluted form. But little attempt was made to find it. Then in 1964 workers at the Bell Telephone Laboratories in New Jersey found, by accident, that the Earth is being bombarded by a stream of long-wave photons apparently coming equally from all directions in space. These, it was theorized, must be the stretched-out descendants of the Bang's compressed photons. The Bang was calculated to have occurred between 13 billion and 20 billion years ago, depending on how one calculated the rate of galactic expansion. Later calculations on the age of stars in the Milky Way galaxy (noted above) seem to move the date more towards the 20 billion point, a long period of time, certainly, but still finite.[11]

By the early 1970s quite a bit was known of elementary particles and how they would act in varying degrees of agitation. By theoretically putting the expanding galaxies into reverse motion – like a film running backward – astrophysicists arrived at the surprising result that all the elementary particles of the universe were produced within three or four minutes (a view vividly projected in Steven Weinberg's 1977 classic, *The First Three Minutes*). This was possible because particles interact in billionths of billionths of a second and because these early particles were in the kind of highly compressed state that would produce unusual intensity of interaction. This interaction was perhaps heightened further by "phase transitions," that is to say, "dramatic shifts from one state, stable at a high temperature, to another state, stable at another temperature," a phenomenon seen as parallel to "three states of water – a gas phase, a liquid phase, and a solid phase." In the transitions from from one phase to another unusually intense activity takes place. As a result, in the first split second of the "three minutes" there might have occurred a "sudden burst of super-expansion" in which gravity for a time acted as an expansive force.[12]

What the particles were in the first fraction of a second is unknown but in the next fraction, it has been calculated, there exploded a "sea of hot [i.e., agitated] photons." These united to form leptons, quarks, anti-leptons and anti-quarks that promptly killed each other off. This annihilation was prevented from being total, it is theorized, because there were, perhaps by chance, perhaps because of an unknown originating factor, slightly more leptons and quarks than anti-leptons and anti-quarks. As expansion continued and the photon soup thinned out, the quarks were able to unite into protons and neutrons, and the protons, neutrons and electrons (leptons) multiplied at an enormous rate – in the midst of a "radiant maelstrom" of photons and neutrinos. As hydrogen consists of one proton with one electron circling it, the protons constituted potential hydrogen nuclei. As the whole mass expanded, the "maelstrom" of photons and neutrinos thinned out still further and the protons and neutrons were able to clash with enough power to unite and form nuclei for deuterium (one proton and one neutron) and for helium (two protons and two neutrons). This, it has been calculated, is as far as nuclear development could go in the brief time-span and rapid expansion of the Big Bang. The universe, then, in the period immediately after its origin was essentially a compact mass of particles, without stars or galaxies or even atoms. It is, in fact, calculated that it was at least another 300,000 years before atoms, with electrons circling the nuclei, could have formed. Galaxies, it is calculated, did not emerge until 2-3 million years later.[13]

In the early fractions of a second in the Big Bang when gravity is thought to have reversed itself, the other three interactions, the electro-magnetic and the strong weak nuclear reactions were, it is theorized, combined in one. This suggestion is made likely by the fact that, as we have noted, the electro-magnetic interaction and the weak nuclear one can unite in accelerator agitation to form the electro-weak interaction. Calculation indicates that the strong nuclear interaction, which holds the atomic nuclei together, would join them at a more extreme point of agitation (beyond present technology to produce). As interactions are interactions between particles, they would naturally change as the character and relationship of the particles changed (with different "tennis balls" of intermediate particles emerging). This aspect of the modern Big Bang theory, then, like its other aspects, fits with the known characteristics of elementary particles. We should note also, as non-Marxist scientists are unaware, that the theory reflects the general dialectical aspects of matter as seen in its later stages: development by the conflict of opposite entities spurring quantitative to qualitative changes, with nodal points – "phase transitions" – of sudden "leaps" of development. The theory gives the dialectical materialist picture of an evolving universe a thrilling new dimension.

In recent years some aspects of the Big Bang theory have been challenged. For instance, it appears that some galactic superclusters are too large to have been formed in the 15-20 billion year span allowed by Big Bang advocates, and some galaxies seem to be further away and moving faster than present Big Bang calculation allow. Some Big Bang theorists projected a basically uniform universe from a uniform Bang, but, as we have seen, observations reveal a universe in which the galaxies are irregularly placed in space, occupying but "a small fraction of the volume available." And galactic superclusters are dragging other clusters towards them. The universe, R. Brent Tully comments, no longer appears "homogeneous" but "lumpy." Finally, the actions of the galaxies as determined by gravity may indicate that some 90 percent of the matter of the universe is "dark," i.e., it does not emit photons. As the gravitational pull of the Milky Way is much greater than can be accounted for by its visible matter, it is theorized that it is surrounded by an immense "halo" of dark matter which contains ten times the mass of the visible matter. And this is presumably true of other galaxies also. What the so-called dark matter consists of is under debate. Some, in fact, doubt that it exists.[14]

Although some still question the Big Bang theory, recent satellite evidence confirms a sudden, expansive origin for the universe, and cosmologists seem largely agreed that the theory requires only modification. There is apparently no serious questioning of the evidence that the galactic clusters are in general moving further away from each other in spite of gravitational pulls between them. Such expansion, some critics argue, does not necessarily imply a Big Bang origin but they seem unable to point to any force that would account for it. Hydrogen cannot be made in stars, but it could have originated in Big Bang conditions, and the present proportions in the universe, of about 3/4 hydrogen, about 1/4 helium and a trace of deuterium are "remarkably close" to what would have emerged from such conditions. Similarly, even if the universal microwave radiation is not entirely even, it seems generally consistent with Big Bang projections. Indeed, cosmologists are now hoping to find minor irregularities in it, for it is difficult to perceive how a totally uniform primordial mass could have produced a "lumpy" universe or galaxies. Nevertheless galaxies were produced, along with the stars in them. Studies of a galaxy 13 billion light years away – which means it is 13 billion years old – indicates that in the early universe star-making was proceding at 10 million times the present rate! Similar studies show stars evolving in response to conflictive forces: "great conglomerations of stars were undergoing major structural change through collisions with each other, galactic merging and other turbulent forces."[15]

If in the Big Bang no elements more complex than helium, with but two protons and neutrons, were formed, where did the others come from? The answer, first projected in the later 1950s, is from "cooking" in stars for the less complex ones before they disintegrate and in the "supernova" explosions of large stars for the more complex ones. This latter theory was dramatically supported when such an explosion was recorded by sophisticated equipment in February 1987. In this explosion a star in our galaxy 20 times more massive than the Sun collapsed in seconds "into a sphere about a dozen miles in diameter," a sphere of neutrons formed by the merging of protons and electrons (whose positive and negative charges cancelled each other out). A gamma ray detector on a satellite logged "gamma ray energies" that "were precisely those expected when radioactive cobalt slowly decays into iron." These detections demonstrated that "supernovae are the breeding ground for elements from iron to uranium and that such cataclysmic deaths planted the seeds for the birth of life on Earth." Our Sun fortunately is a comparatively recent star so that it and its planets are the heirs to millions of years of such explosions. The oxygen we breathe, the carbon that holds us together, the calcium in our bones, were formed inside stars and scattered into space along with the iron in our blood. Also, without iron there would have been no industrialization.[16]

The evolution of the universe, then, reached a new stage with the formation and explosive deaths of giant stars. As we have seen, astrophysicists, basing themselves on what is known of elementary particles in different states, can trace the universe back to a "sea of hot photons." We are not, then, looking for an act of "creation" for "the universe," but for the origin of a group of agitated photons. This origin, it has been speculated, perhaps lay in a single "hot seed" which was only "one trillionth the size of a proton." Where, however, did such a "seed" come from? Perhaps, it has been suggested, by an accidental "vacuum fluctuation" of "nothing." Nothingness, it is speculated, existed for an indeterminable time, but it was fundamentally "unstable" so that sooner or later a "singularity" would occur. This speculation, which is being seriously considered by a number of scientists, and hence, worthy of note, clearly runs into problems. If the "seed" was hot, how did it get hot? What we call heat is actually an agitated interaction of particles (or atoms or molecules). So the seed must either have been interactive with other "seeds" or had interactive component parts. If there was a "singularity," then, it must have been one of interactive units and these units can hardly have been anything else than a form of matter. But it seems difficult to conceive of any unit of matter less complex than a photon. An isolated "seed" a trillion times smaller than a proton is pure fantasy, without basis in particle physics. Whatever first arose – if the phrase has any meaning – must have been plural and not singular.[17]

Some also speculate that before there was matter – in particle form – there was "space-time." But the fact that time is measured in billionths of billionths of a second in particle reactions, and space in such reactions is similarly miniscule, does not mean that either space or time can exist independently of particle reactions. All three seem to exist in a dialectical interchange, in which all are necessary but matter is basic.[18]

It does not seem possible, then, in our present state of knowledge, to push the evolution of the universe back beyond the "sea of hot photons" which coalesced to form the particles that later combined into atoms.

The picture now emerging of the universe further reveals the basic correctness of materialism. The universe, Lucretius tells us (partly echoing his Greek predecessors), arose by natural causes from a "hurricane" of "atoms," from whose "disharmony" there sprung "conflict" out of which "the stars" and the Sun arose. All the variety of nature was determined by "in what combinations and positions the same elements occur." There is neither divine creation nor supernatural direction. The details are wrong, of course, but the general concepts are right, in contrast to the fantasies of the idealists. The more sophisticated views of Engels, reflecting, of course, the new science of the 19th century, are closer than those of Lucretius to the general picture being revealed today: a universe of blind, natural forces (and nothing else) where qualitative change arises from quantitative and arrangement changes (and, again, from nothing else), changes themselves based, directly or ultimately, on the oppositeness of elementary particles.[19]

As I have noted, it was first theorized and only later confirmed, by "atom-smasher" experiments, that the electromagnetic and the weak nuclear interactions were basically one, improbable though this had seemed, for each is dependent on a different elementary particle. One of the theorists, Steven Weinberg, on hearing of the experimental confirmation, commented: "There's now such a sense of confidence. It doesn't seem as if we are making it up as we go along." Even the scientists involved, then, have some sense of unreality about the extraordinary things they are uncovering. It was, it has been suggested, this attitude that kept researchers from looking for the left-over radiation from the Big Bang. Although parts of the picture will prove to be wrong and new perspectives will emerge, enough basic data now fits together to indicate that the present picture is getting close to the reality, and the reality is close to us. There really was a Big Bang and its photons are daily showering on the Earth. The atoms that compose us originated in the fiery heart of immense stars. The particles that compose these atoms originated in the Big Bang.[20]

"The more the universe seems comprehensible," Weinberg writes in the conclusion of *The First Three Minutes,* "the more it also seems pointless." He sees the universe as "overwhelmingly hostile." We can see the universe as "pointless," however, only if we are looking for a "point," that is to say, for something that it is not in the nature of matter to possess; in short, if we are looking for God, or, more elegantly, for Purpose. But although the universe is inconceivably vast it is everywhere essentially the same and gives no indication of Purpose or God in its mindless mechanistic repetitions. On the other hand, it is neither hostile nor non-hostile – any more than is a chair or a windowpane. True, the workings of matter will destroy Earth and us sooner or later but they also gave us life. What we witness in the evolution of the universe is the evolution of matter to the point at which living matter could emerge from it.

Living Matter and Life

As we look at a space photograph of Earth, with its forests, oceans and swirling clouds, it seems alive – in dramatic contrast to the dead, pocked plains of Mars, the blazing vapors of Venus or the still ice of Jupiter.

Life on Earth consists of bacteria, plants and animals in an interactive complex. Plants could not live without bacteria and animals could not live without both plants and bacteria, and bacteria need one or both for food. The animals range from insects (800,000 species), fish (23,000 species), reptiles (5,000 species) and birds (8,600 species) to mammals (4,500 species, of which primates, including humans, form a subdivision).

Engels hailed as one of the milestones of science the discovery that the substances of life exist in interconnected "cells," which are similar in structure to a plum or peach, with a central core (the "nucleus"), a layer of fruit (the "cytoplasm"), and a skin (the "membrane"). He also believed, reflecting the science of his day that the only substances of life were proteins, fats and carbohydrates. And of these he saw the proteins as the basic and essential generative substance. Nothing was then known of the "nucleic acids" which shape the protein base. Nor was it known that proteins consisted of the molecules of simple organic compounds, the amino acids, linked together in long chains, looped and shaped in various ways.

The differences between the elements in general, between, say, oxygen and iron, arise, as we have seen, from the number and arrangements of their atomic particles. So, too, with the chain molecules of life. Protein consists of hydrogen atoms, with one proton and one electron, carbon with six protons and electrons, nitrogen seven, oxygen eight, and sulphur sixteen. The electrons are arranged, as in all atoms, in various orbits, those in the outer orbits able to combine with the outer electrons of other atoms in negative-positive combinations and thus unite the atoms into molecules and the molecules into chains. Whereas in atomic matter, with its numerous elements, molecules are comparatively simple structures, consisting of a dozen or fewer atoms, the protein chain molecules (or macromolecules) contain between 18,000 and 10,000,000 atoms, held together by carbon, the great binder among the elements. As Francis Crick notes, not knowing that he was looking at a general dialectical process in nature, "How each protein acts depends on its exact three dimensional structure." Qualitative change – producing function – then, depends on the number and arrangement of the protein's constituent chain molecules. We have moved from one level of matter to another, from atomic matter to living matter, from a stone to a dog, essentially through combinations of the same units.[21]

Although their functions are so different, the components of proteins – amino acids – and the nucleic acids are very much the same. Like proteins, nucleic acid molecules consist of hydrogen, carbon, nitrogen and oxygen atoms, but in the place of sulphur as a fifth constituent, we find phosphorus, with fifteen protons and electrons to sulphur's sixteen. The names of the two main kinds of nucleic acid chain molecules, ribose and deoxyribose, or RNA and DNA, indicate that the sugar (ribose) molecules of one contains one less ("de") oxygen atom than the other. And that is the only quantitative difference between them. Both the difference in function between the proteins and the nucleic acids and that between the different types of nucleic acids, then, arise from the number or arrangement of the molecular groupings of six kinds of atoms, which themselves differ basically only in the number and arrangements of three particles – protons, neutrons and electrons.

The number and arrangement of units is crucial also in the process of heredity. The nucleic acid chain molecules in all the cells, including the sex cells, are, it was discovered in 1953 by Francis Crick and John D. Watson, arranged in double spiral formations, rather like intertwining spiral staircases (arcanely called a "double helix"). There their molecules flutter out like filaments into the spiral staircase well. In the well, molecular filaments intermingle in a complex of combinations. In the sex cells these combinations shape those of the protein molecules that in turn shape the growing embryo. And when the embryo develops, is born and can itself become a parent, the nucleic acid molecules in its sex cells are transmitted again to supply half the base for a new embryo, half from each parent. When the nucleic acid molecules are directly changed, as by X-rays or gene-splicing, these changes, too, are transmitted and affect all future generations. Heredity, then, arises from the arrangement of the nucleic acids in the sex cells. The sex cells are spurred to interaction by their inherent opposite – negative-positive – atomic charges.

The precise way in which the nucleic acid molecules mold the formations of protein molecules is now becoming clearer. The first stage (DNA) exists only in the cell's inner core, the nucleus. In what seems to be a general process in living matter, it contains many more molecules than are needed for the production of the offspring. In fact, 98 percent of these molecules are eliminated by "suppressor" molecules until only the less than 10 percent needed for the offspring are left. This trimmed down nucleic acid is called "messenger" ribose nucleic acid (RNA).

Whereas the DNA molecules remain tucked within the nucleus, the RNA molecules move out into the surrounding cytoplasm, where after further rearrangement, they latch on to bunches of free-floating amino acid molecules and shape them into the jigsawlike protein patterns from which the embryo develops, rather like a magnet shaping iron filings.

"How," some tend to ask on first learning of these things "do the suppressor molecules 'know' how to act in this way?" The answer, of course, is that they "know" nothing. That their action is automatic is shown in the mechanical and roundabout way in which they operate, producing qualitative change by first increasing and then decreasing the number of nucleic acid molecules. Complexity, as with a computer, arises from two simple basic units, positive and negative atomic charges. And it all takes the particular forms that it does because it was molded by trillions of actions and counteractions in the long, volatile chemistry of evolution.

Atomic matter, as we have seen, arose from particle matter through combinations of the particles, and although it still contains the particles, its essential functional unit is the atom as a whole and not the particles. Similarly, although living matter arose from atomic matter through combinations of its atomic packages (molecules) into the still larger packages of the chain molecules, it is in these packages that, even though it still contains atoms, its viable essence lies. In short, life is the functioning of matter in its chain molecule form, the functioning arising from the size and structure of the chains and nothing else.

Although materialist philosophers from Lucretius to Diderot had argued that there was no "creation" of life and that it had emerged from matter they had no specific concept of how this had happened. They did not know that atoms differ from each other or that certain atoms form living matter, particularly those of hydrogen, nitrogen, oxygen, carbon, phosphorus and sulphur. Nor did they know that some of these atoms also exist on other planets, for instance, in the ammonia (nitrogen and hydrogen) clouds of Jupiter or the carbon dioxide (carbon and oxygen) atmosphere of Venus. On comparatively nearby "interstellar clouds" carbon, hydrogen, oxygen, nitrogen and sulphur have been detected, sometimes united in the gas cyanogen (carbon and nitrogen). We find in the present oceans not only nitrogen, hydrogen and oxygen, but also sulphur and phosphorus; and these, it is indicated, existed in earlier oceans also. When these elements are put together in a laboratory and subjected to photon and electron bombardment – duplicating the sun and lightning of the early Earth – protein molecules and elements of the nucleic acids emerge. Here then, is a bridge from atomic matter to the first elements of life on Earth, a bridge that led first to the chain molecules and then to the cell. Cell formation involves a spiraling intermingling of these molecules, all the stages of which are not yet known. It was theorized that a protein enzyme would be needed to spark the process. But how could protein function without nucleic acid? The problem was apparently solved when it was found that RNA can "cut itself, thereby altering the material it produces, an operation that previously had been thought to need an enzyme." RNA, that is to say, acts like a protein catalyst, setting the molecules in motion and speeding them up.[22]

This biological perspective is generally supported by the geological evidence. The Earth was solid, it is now believed, by 4.6 billion years ago. The first evidence of life occurs about one billion years later – in the form of microscopic one-celled creatures that lacked a nucleus but apparently produced oxygen. An immense two billion years then seem to have passed before purely chance interactions produced one-celled creatures with a nucleus and opened the way to higher forms of life, which then took another billion. It is now theorized that some early activity might have been triggered by volcanic vents in the ocean floors as well as in sun-baked sea-shallows, where it perhaps formed a kind of chain-molecule "soup." But no matter what the particular spurs to life, the classic materialist argument that it arose from matter and nothing else has begun to be substantiated.

It became apparent in the 19th century, as Engels noted, that life on Earth was powered by the Sun. In the 20th century we can see more exactly how this takes place. The cell is the basic life-unit of plants as well as of animals, and in plants it has the same structure of inner nucleus and outer cytoplasm (with various smaller bodies floating in it) surrounded by a membrane. In one of the smaller bodies in the cytoplasm of certain plants is a grouping of hydrogen, oxygen, and nitrogen atoms linked by carbon to a magnesium atom, the total forming the molecule of the green pigment chlorophyll. When the photons of light rain down on the chlorophyll molecules in the leaves, etc., the electrons of their constituent atoms shoot rapidly from one end to the other of the molecule. The photon shower, in other words, sets up a stream of electricity in the plant's cells – the universal process of photons prodding electrons into action, and vice versa. One effect of this activity is the production of the molecules of sugar (carbon, hydrogen and oxygen) which, especially in its form of cellulose, is an essential part of the plant's structure. Although all the numerous and minute steps of this conversion process in plants are not yet known, the general outline is emerging: "Carbon dioxide is taken in from the air, reduced to carbon, combined with water to form a carbohydrate – sugar, starch or cellulose – and the extra oxygen restored to the air." (When years later the fossilized plant is burned in the form of coal, oil or natural gas, the stored-up carbon is restored to the atmosphere in the form of carbon dioxide, the main gas behind the growing disaster of global warming.)[23]

Thus the electrons of plants under photon bombardment from the sun turn, then, into a tool (electricity) to make a biologically functioning energy source, namely sugar. When animals eat the plants, this bio-electric energy is transferred to the animals. For this and other reasons, a British science writer commented that the "life force" has turned out to be "simply the electrical force." Not really "simply," of course for naked electricity would kill life, but an electrical force muted by the massive chain molecules of living matter. A physical force that would destroy life is thus converted by a quantitative process into the essence of life.[24]

Animals, from insects to mammals, reproduce by uniting male and female sex cells, usually – fish and amphibians form an exception – through copulation. In the human female the sex cell, the ovum, is 1/200 of an inch long, the male sex cell, the spermatozoon, is 80,000 times smaller. One ejaculation contains between 3 million and 10 million spermatozoa, only one of which, in fertilization, penetrates the ovum. The fertilized ovum contains in its nucleus a double helix of DNA chain molecules, which if stretched out would be 1/4 of an inch long. In molecular terms, of course, 1/4 of an inch is enormous; and the chain is extremely thin: "If the cell's nucleus were to be magnified 100 times, it would be about the size of a pinpoint just visible to the eye. The DNA folded up in that nucleus would be the length of a football field!"[25]

In a fertilized ovum half the DNA chain molecules come from the mother, half from the father, for in both testes and ovaries the sex cells' nucleic acid chain molecules are divided in half. If body cells are artificially made to reproduce, they create a simple duplicate – clone – of the individual containing them, but when the fertilized ovum reproduces it creates a unique offspring from the intermingling of two different nucleic acid chain molecule groupings, half from the ovum, half from the sperm.

The general process by which the fertilized ovum, a single cell, "differentiates" into different cells – those for skin, bones, nerves and so on – has become somewhat clearer in recent decades:

> The cells of the liver or of the brain or of muscle all have the same genes in their nuclei. Experiments have proved the truth of this statement: a frog's egg whose nucleus has been artificially replaced with the nucleus taken from a skin or liver cell gives rise to a normal individual. . . . If a living cell is removed from its natural environment in the body and placed in a culture fluid in which it can grow and multiply, it often loses the characteristics of the organ it came from and becomes more or less similar to the undifferentiated cells from the early embryo.[26]

Liver cells in a developing embryo, for example, have the potential to be skin cells or brain cells or sex cells. But if all cells have these immense potentials, why is life not a wildly chaotic profusion? The answer is that "suppressor" molecules eliminate most of the cell's molecules, so that some cells lose all their molecules except those that will form liver and others lose all except those that will produce skin. There is thus no sudden creation but, once more, a long mechanical, roundabout process of elimination. The suppressor molecules do this not, as one would sometimes gather, by animistic "clippers," but by atomic electrical actions and reactions which we might envisage as complexes of tiny lightninglike flashes. And once more the new is produced by quantitative action, this time not by adding but by taking away. This process is often presented as a "harmonious series of events," in which certain cells "are preprogrammed to die at a certain point," but experiments by Dr. James Michaelson of Harvard indicates that they die off in sharp competition. Thus behind the quantity-to-quality change there stands, as ever, conflictive forces. Life conflict begins in the embryo.

Such, then, is one general process at work. But what of the processes that change a mass of cells into an organism?

It had been known that a certain group of proteins controlled growth in fruit flies. When the fruit fly was mutated by X-rays the proteins also gave it a "bristly appearance." Hence the group was dubbed "hedgehog" proteins. In January 1994 it was announced that these proteins also determine the shaping and differentiation of the embryo in mammals and other vertebrates. However as the molecular structure of the proteins has not yet been unravelled, it is not yet clear exactly how they work. But, it is clear that they do work – and biology is on the threshold of a new major advance.[27]

During the normal functioning of a cell in maintaining and renewing itself, stringlike bodies containing the gene packets of the nucleic acid chain molecules are faintly visible under the microscope. When the fertilized ovum begins to reproduce, these bodies, the chromosomes, become prominent, and begin to take over specific tasks, including the determination of sex. There are in all the body cells of a male animal – again from lobsters to humans – two chromosomes that determine sex, an x and a y, and in a female two x's. When the genetic material is split in two (in ovaries and testes) the male sex cells get either one x or one y and all the female sex cells one x. If the one male sex cell (spermatazoon) that penetrates the female ovum has an x chromosome, the offspring is female, if it has a y, the offspring is male. This comparatively simple mechanism not only divides all animal species into male and female but has made the evolution of higher forms of life possible, including us.[28]

To sum up, it is apparent that living matter is a form of matter, arising directly from atomic matter. Its atoms react as they do in atomic matter, say in a stone or a gas. Because they are clumped in large molecules they react in more complex patterns and it is this – and only this – that makes the difference. Although the basic mechanism is the same in both matter and living matter, namely negative and positive atomic charges, size and pattern (arrangement) and not the atomic mechanism directly determine function. Like molecules, chain molecules do not act as random collections of atoms but as units. The basic materialist outlook, then as presented by Lucretius, for example, is still generally valid but it is now apparent that matter has different levels of complexity and these have different viable units, from particles to chain molecules. It is apparent also that change takes place on all levels – from light to life – by the same general mechanism, namely by additions, subtractions and arrangements of various units spurred by the interactions of entities that are either directly or ultimately oppositional. And to these factors we sometimes have to add, as with retinoic acid, "timing." The time at which certain reactions take place can have a basic effect in biological and other processes. Again, as with arrangement, this seems to be an essentially quantitative factor.

The Ways of Evolution

By the mid-19th century, numerous fossils of plants and animals had been uncovered in rock layers and showed some lines of progression from simpler to more complex forms. To this geological story Charles Darwin added his own varied observations of animal life in the wild – including the Galapagos Islands – and came up with the theory whose nature is indicated in his title: *On the Origin of Species by Means of Natural Selection, or the Preservation of Favoured Races in the Struggle for Life* (1859). If, Darwin reasoned, animal breeders can produce particular strains by deliberate selection, could not the same thing happen in nature unintentionally by "natural selection," slowly, over the eons, in the endless rough-and-tumble of the "struggle for life"? "Under these circumstances," he wrote, "favourable variations would tend to be preserved, and unfavourable ones to be destroyed."[29]

Along with "natural selection" Darwin included "sexual selection." This affected male animals in particular, for, Darwin argued, the males in their "struggle" to impregnate females developed particular characteristics designed to attract the females, such as bright plumage and long tails among birds. This theory has now been tested in experiments. They indicate that such characteristics as "the peacock's technicolor tail and the bullfrog's booming moonlight sonatas evolved for no other purpose than to allow males to curry favor with females."[30]

When Darwin (and Engels) wrote, the geological record was sketchy and the science of genetics had not been born.

After crude one-celled life forms, lacking a nucleus, first came into being some 3.6 billion years ago, why did it take the immense span of three billion years to evolve more complex forms? The answer seems to be that such change could take place only after a certain amount of oxygen had been created by the action of bacteria and sea plants. About 600 million years ago multicelled creatures first appear; and from there on the record reveals increasing complexity (amid ups and downs): by 500 million years ago, sea-worms and sponges; by 450, clams and star fish; by 400, the first land plants; by 350, amphibians, primitive fish and more plants; by 300, reptiles (including dinosaurs) and more complex fish; by 150, birds and small mammals; by 60, large mammals and small (ancestral) primates – "prosimians." Along with these changes in species, the species themselves split up. About 30 million years ago, apes split off from monkeys; about 25 million years ago; dogs split off from racoons.[31]

There is clearly, as George Gaylord Simpson noted in his classic *The Meaning of Evolution,* "a tendency to spread and fill the earth with life whenever and however this chanced to become possible" – even through cracks in a sidewalk. This "tendency" must be ultimately based, as Darwin did not know, in the sharp driving spurs of electrically conflicting chain molecules. The story of evolution is not, as we would gather from some commentators, one of benign unfolding but of blindly clashing forces.[32]

Along with complexity the record reveals generally increasing variety as "the struggle for life" reacted to environmental changes. Dinosaurs developed forms for land, air (Pterodactyl) and sea (Ichthyosaurs). There are sea mammals, whales and dolphins; air mammals, bats; and underground mammals, moles. Birds vary from sparrows to swans, from penguins to ostriches – all, again, with features suited to their natural environments. In just one film on life forms in but one area, the Andes, we see that llamas living in high altitudes have more than the average number of red blood cells and that frogs that live under water by breathing through their skins develop "baggy pants" of extra skin to give more breathing surface. In but a few nature films, we can see several hundred such adaptations of incredible variety. In short, small biological – potentially evolutionary – changes are taking place in all species all the time. We can see their results in evolutionary histories. The first horses (Eohippus), living in forests, were small, with doglike paws instead of hooves, and teeth suited for munching leaves; the later horse (Equus), a plains-dweller, had hooves and grass-grazing teeth. Camels and horses evolved differently in North America and South America, monkeys differently in South America and Africa. Conversely, life forms that are adjusted to an unchanging environment change very little. Oysters have been about the same for 200 million years, bats for 60 million, bees for 80 million.[33]

Of the fact of evolution then, there is no question. The geological record reveals indisputably that it took place. The question is how? And here, although some basic answers have appeared, problems still remain. Both Darwin and Engels believed that characteristics acquired by the parents prior to the birth of the offspring, such as the muscles of an athlete, could be inherited. This seemed to present a relatively simple mechanism for evolutionary change. However, as we have seen, the mechanism of heredity exists only in the nucleic acid molecules in the sex cells and change in heredity comes only from changes in these molecules. We are, then, left with the problem of discovering how apparently encapsulated genetic processes can bring about bodily change and integrate it with the "struggle for life."

When today we examine the course of evolution of any particular species, the 19th century view of steadily mounting development is seen to be wrong. As noted by George Gaylord Simpson in regard to the horse:

> In the particular lineage from Eohippus to Equus, general foot mechanics became first more complex, then simpler. The number of toes did not change at even pace from four (in the forefoot) to one, but changed in two spurts, first from four to three, then much later from three to one, each rapid adjustment to the new sort of foot and to changes in the weights of the animals.[34]

The picture from the geological record, then, is that of some lines of evolution advancing in complexity but in a kind of hit-or-miss way and of others dying out or hardly developing at all. In fact, "about ninety-nine percent of all life forms that have ever existed on Earth are now extinct." Clearly there is no Tennysonian "one increasing purpose" visible in evolution, no divine guiding hand, no Aristotelian "end." Nothing but mechanical processes.

If changes in inheritance can take place only by changes in the nucleic acid molecules, changes that are then transmitted to future generations, what can bring about such change? It should be noted first that the nucleic acid segments do not sit like blobs in fixed positions in chromosomes but, to quote John Gribbin, are in "a state of dynamic change" and occasionally move between chromosomes.

It has been demonstrated experimentally that changes can be made in the nucleic acid molecules of sperm, ovum or fertilized ovum by photon and neutron showers, by other molecules (chemicals), viruses and heat. Thus the genetic material of fruit flies bombarded by X-rays (photons) in a laboratory have produced offspring with new wing shapes or colors, and these changes have then been transmitted to offspring and offspring of offspring. Apparently, however, the main source of mutations in nature lies in none of these but in accidental matchings of molecules in the multi-billion exchanges in the immensely long and volatile nucleic acid double helix windings. Although such errors are comparatively few, they are large in total numbers. Genetic change, then, is a purely chance affair.[36]

Genetic change alone, then, cannot account for evolution but only lay the basis for it. As the "rate of mutation" is "roughly constant" in all species, there must have been as many genetic changes in bat nucleic acid molecular chains as in those of horses, but horses have evolved and bats have not. Bats in their caves have had a stable and unchanging environment. What bodily changes took place in some were not needed for the survival of the whole and were lost in the whirl of generations. But horses were apparently forced by climatic change from forest to plains. In this situation certain genetic changes – and the animals they produced – survived and in time led by a long, erratic route to a "new sort of foot." We do not, then, have a closed system but one subject to outside forces; for instance, those brought about by continental drift or recurrent ice ages. "Darwin," Francis Crick writes, "believed that the *appearance* of design is due to the process of natural selection," of struggle within and between species. Now, however, it has become clear that natural selection does not work alone. Geological and ecological processes seem to shape basic directions for the biological "struggle for life."[37]

Let us return, for an example, to our high altitude llamas and their extra red blood cells. How could they develop such cells? And if one did, how could they be inherited by offspring? The answer now favored is that among the immense number of small genetic changes constantly arising in any animal population, there were some – most probably arising from chance mismatching in the double helix interminglings – that over the generations produced some llamas with extra red blood cells. In "the struggle for life" in high Andes' conditions, this strain developed while others went under.

Although, then, genetic research has robbed us of the simple, direct mechanism of the inheritance of acquired characters – which, in fact, had it existed, would have not produced evolution but chaos – it has provided a complex but consistently materialist base for evolutionary process. Once more, as, for instance, with embryo activity, we see nature acting in a purely bio-mechanical manner, its processes a combination of chance and the directions arising from chance.

One problem that is being debated is how a succession of minute changes could bring about not only large but highly coordinated changes – in bones, blood vessels, muscles, nerves and brain – such as those involved as one set of reptiles evolved into birds and another into mammals. Or the change within one species of mammals, exemplified by apes and humans today, that produced uprightness. In approaching this problem we should note, first, that theories of sudden "macrochanges," which some proposed, do not hold up. They would, as various scientists have pointed out, almost certainly be lethal. Nor is there any example in nature of such change. Even major "spurts" in development in any field are not at all of this magnitude; and they apparently arise only after long accumulations of smaller changes. This, in fact, seems to be the only way in which living matter can act. The genetic complex for any species is formed not by sudden "creation" but by a laborious, roundabout and computerlike process of making and destroying, often to produce simple change. We are back, then, with what Sewell Wright calls "trial and error processes at the level of small random changes." In short, we face the same general process as at the beginning of evolution when the spiraling interactions of the macromolecules of the proteins and nucleic acids led to cell formation, with its complex coordinations. We should note also that a small genetic change can produce a large bodily change and that some genetic segments regulate processes that control structures. Such changes clearly have large ramifications. Although the genetic difference between the great apes and humans is apparently but one percent, this one percent obviously entails great bodily (including neurological) change.[38]

Although the specifics of the interlockings that produce massive changes in evolution can finally be determined only by further research, it should be noted that the problem involved is not confined to natural evolution. We have a similar problem in social evolution where such complex coordinated changes – economic, social, political, cultural – as those involved as capitalism emerged from feudalism took place as the result of a number of smaller changes that endlessly interacted and in time produced major changes. If we view this phenomenon mechanistically – in hen-and-egg logic – and not dialectically, it appears impossible. Political change could not have come about without economic change and economic change could not have come about without political change; so, too, with social and cultural change. The solution appears only when we realize that these things did not develop in isolation but in mingling back-and-forth movements of basic and derivative elements on various levels.

Human Evolution

One of the most dramatic discoveries in the history of science was made in 1977 in East Africa (Tanzania): two sets of fossil footprints preserved in volcanic ash some 3.7 million years ago which look very like human footprints left on a beach today. The subhumans that made them, and were trapped by a volcanic eruption, walked upright and were apparently similar to those represented by the skeletal remains of

"Lucy" discovered some 700 miles to the north, in Ethiopia, in 1974. "Lucy," a female about $3^1/2$ feet tall and weighing about 60 pounds, also walked upright. She had humanlike hands but her jaw, skull and teeth look rather apelike. The males of the species, found some 16 years later, were almost twice as heavy as the females – as with gorillas, orangutans but not chimpanzies. Human males today are on the average 15-20 percent heavier than females. The Lucy subhumans seem to be of the same general nature as those of the first subhumans (hominids), who inhabited the east African plains from perhaps five million to one million years ago, named (or misnamed) "australopithecines" (southern apes). One find ranged up to five feet tall, weighed up to 150 pounds and had a brain capacity of 380-450 cubic centimeters. (Chimpanzees, of comparable size, have brain capacities of 300-400.) Microscopic examinations of their teeth indicate that their diet was similar to that of the chimpanzees, mostly vegetarian with emphasis on fruit. No stone tools have been discovered with the remains. In short, these first subhumans seem to have been essentially upright apes but perhaps with somewhat higher intelligence. They lived, however, not, as did the apes, in the forests but in the plains, to which they were apparently driven, like the horse, by climactic change. The indication at present is that early subhumans evolved directly from the apes (who have been around for some 20 million years) some 4-5 million years ago.[39]

The record of pre-human evolution beyond "Lucy" has filled up somewhat in recent decades but is still sketchy. By 2.5 million years ago there had evolved what looks from the skeletal remains to be a more developed form of australopithecine – *homo habilis* – about 4 to $4^1/2$ feet tall, 65 to 110 pounds, with a comparatively large brain capacity of about 640 cubic centimeters. They apparently lasted some 900,000 years. Stone tools have been found with *homo habilis* remains and there is some indication of shelter building. At about the same time, the first traces of a larger, more "robust" australopithecine appear, traces that vanish at about one million years ago. About 1.7 million years ago, the first traces of a considerably higher form – *homo erectus* – appear in Africa and subsequently in a wide area, from Europe to China and Java. Cave remains in Africa reveal the use of fire, at least one million years ago. Skull remains indicate a brain capacity ranging from 900 to 1,100 cubic centimeters, the upper level coming within the human range – 1,000 to 1,800. The (extensive) Chinese remains show that these *homo erectus* people also had fire and hunted large beasts – about 600,000 years ago. What are recognized by some experts to be fully human people and regarded by others as still slightly subhuman, the Neanderthals, first appear about 150,000 years ago and cover an area from Europe to West Asia. Generally recognized fully human remains have been dated at about 100,000 years ago, tentatively in Africa, more firmly in the eastern Mediterranean.[40]

As we look over this body of evidence, it appears that the early evolution of humans followed a similar pattern to that of the horse. The general features of both developed in forest conditions, the soft-pawed early ancestors of the horse browsing on plants and leaves, those of humans living largely in trees. The specific features which later marked both, including uprightness in humans and hooves in the horse, although perhaps, it has been suggested, developing to one or another degree in a minority population in the forests, apparently first became general on the plains. In both species further early evolution seems to have been connected with plains living, the horse as grass eater and swift runner, the human as hunter and food gatherer, the change in both species apparently forced by climatic changes that destroyed forest areas and spread grasslands. As with the evolution of the horse, that of humans went through a hit-or-miss process in which some lines, such as the "robust" australopithecines, deadended; and for long periods there were apparently few changes. Some australopithecines remained about the same for three million years. As Simpson noted of the horse, however, there appear to have been some comparatively "rapid transitions," for instance in the emergence of *homo erectus.* When we look at a considerable sequence of subhuman and pre-human skulls, such as that presented at the Museum of Natural History in New York in 1984, the *homo erectus* skulls stand out immediately as the first to look more human than apelike. There is some indication that the Australian natives are the direct ancestors of two streams of *homo erectus,* one from Indonesia, the other from China. As the Australian natives are fully human, could this mean that advanced *homo erectus* had a full human potential? If so, humans or virtual humans have been on the earth for more than a million years.[41]

The fossil record, then, indicates that humans evolved in the same ways as other animals. It shows also that Engels' suggestion – via Haeckel – that uprightness came before large brain size is correct. Nor is any special process needed to explain the evolution of the human brain from the ape brain. It must in fact, Stephen Jay Gould suggests, have been less difficult to evolve than uprightness. The fact that we look so different in some ways from the apes was early used to ridicule Darwin. These differences, however, as we noted, are produced by only a one percent deviation in genetic material. It has also been discovered, as John Gribbin notes, that humans bear a stronger resemblance to the ape fetus – "from the shape of our heads to the absence of hair on our bodies" – than to the mature ape. The indication is that basic difference between apes and humans resulted from a "slowing down" in human development which allowed the brain to "grow long past the point where the baby's head would be too big for its mother to give birth and live." This difference could apparently arise from minute changes in the molecules that regulate growth rates. There was of course, no direct evolution of humans from apes but of hominids like the "Lucy" subpeople of some 4 million years ago, perhaps from chimpanzees or from a common ancestor of the chimpanzees and gorillas. The parallels between these subpeople and one kind of chimpanzee, the so-called Pigmy chimp, appear to be very close.[42]

The (necessary) division of the sciences into physical and biological obscures the fact that we are dealing with a continuous story, from the earliest – Big Bang – forms of matter to biological evolution. Nor was there any break as social forces blended with biological ones to form people. New sciences have arisen that throw light on these blendings.

Chapter Five

BRAIN, BODY AND MIND

Although for centuries philosophers speculated about the nature of mind, all they knew of it was what they learned by observing the thinking process in others and in themselves. Earlier philosophers, from Plato to Kant, did not, in fact, know that there was a connection between mind and brain. Even after some elementary facts about the brain began to be discovered, philosophers on the whole tended to ignore them, preferring to deal with abstractionist models of "the mind," and still do so today. However, the comfortable premise that this "mind" is a unique entity which, no matter how much we learn of the brain, is itself forever unknowable, has become increasingly untenable in recent decades in the face of psychological and neurological research. Today we stand, as it were, face to face with the brain, not only when we think but when we walk or fight or talk or make love.

The general dependence of "mind" on brain was becoming apparent to later 19th century scientists. "Thought and consciousness," Engels wrote in 1877, stating the essence of the "modern," evolutionary, materialist position, " . . . are products of the human brain and . . . man himself is a product of Nature." Evolution had provided "the basis- . . . for the pre-history of the human mind, for following all its stages of evolution from the protoplasm, simple and structureless yet responsive to stimuli, of the lower organisms right up to the thinking human brain. Without this pre-history, however, the existence of the thinking human brain remains a miracle."[1] The brain-mind puzzle, that is to say, cannot be solved by philosophical speculations but only by going back to our evolutionary roots and little by little, working upward.

Marx, advancing beyond the current materialism, rejected the photographic, abstractionist "mind" separated from body and life. "The chief defect of all hitherto existing materialism," he argued in *ad Feuerbach,* " . . . is that the thing, reality, sensuousness, is conceived only in the form of the *object* or of *contemplation* but not as *human sensuous activity, practice,* not subjectively." The views inherent in both statements have been supported by 20th century science. Psychological investigations have begun to show how "thought and consciousness" are related to brain responses. Neurological research plus field investigations into the relation of "man" and "Nature" (specifically animal societies) have alike begun to reveal the brain's evolutionary ladders. The implication of Marx's statement that "mind" cannot be considered in isolation from "sensuous activity" has been documented by the discovery of nerve and hormonal connections between brain and body. Lenin, in the dawn of the new physics, saw "the energy of external excitation" being "transformed" into "sensation" and sensation into "consciousness." He argued that "concepts are the highest product of . . . matter." They arise from matter but are not themselves matter. These views, too, have begun to be substantiated as more is learned of the electronic aspect of the brain.[2]

Nevertheless, although Marx, Engels and Lenin were on the right track, neither they nor anyone else in the period covered by their writings could explain how the brain could transform external stimuli into thoughts and emotions. Nothing was then known of the nerve cell (neuron) or of hormones, and little was known either of the structure or the exact evolutionary history of the brain. What has so long seemed a mystery is, as Engels anticipated, solvable in terms of very small steps occurring over many millions of years. And it becomes a study in what Lucretius insightfully called the problem of how the "sentient" arises from the "insentient."

The nerve cell, as Marx and Engels would have been happy to hear, is the functioning cell of both the brain and the nervous system, thus forming an interactive unit of brain and body. Like all cells, the nerve cell – in cats or humans – has a nucleus enclosed in a cytoplasm, with the whole encompassed in a membrane that admits some molecules and keeps others out. Like other cells it generates electricity (from sugar in the blood stream). But unlike other cells it has long strings of its own substance branching out at each end through which the electric current flows. (The current "is only $1/10$ of a volt – more than a thousand times smaller than household electricity" and "it lasts only $1/1000$ of a second.") Like other cells, the nerve cell is largely liquid, a salt water solution. A similar solution surrounds the cell on the outside. This simple pattern, when functioning in interaction with the nerve cell's unique branchlike structure, provides the mechanism for nerve and brain activity. In short, the functional difference between nerve cells and other cells simply lies, as with liver cells or skin cells, in structure (arrangement) of the same basic units.

The salt solution inside the nerve cell, as in all cells, consists almost entirely of potassium and chloride molecules; the solution outside the cell of sodium and chloride but with less chloride. Thus we have sodium chloride (table salt) outside the cell and potassium chloride (another kind of salt) inside. As both sodium and potassium atoms – with, respectively, 11 and 19 electrons – have but one electron in the outer ring, they tend to lose this electron and become positively charged. Chloride, on the other hand, tends to gain one electron and become negatively charged. As there are more chloride atoms inside the cell than outside, the salt solution inside the cell tends to be negatively charged and that on the outside positively charged. It is this volatile imbalance between negatively and positively charged atoms that activates the electrical current (a fluctuating stream of electrons) and sends it through the membrane wall and, in the case of the nerve cell, down its long, filamentlike branchings, which act like an electric wire. It is by means of this current in the nerve cell system that animals and humans are able to move and breathe, smell and remember. The nucleic acids and proteins, for all their extraordinary powers, could not give life above a single cell level, as in the amoeba, without the vital electric spark of the nerve cell.[3]

The nerve cell is the same not only in mammals but in all animals that possess it. The lobster nerve cell has, in fact, served for experimental investigations into the general nature of the nerve cell, including the human. The difference between the nervous system of lobsters and humans does not arise, as one would think, from a new basic substance but from an increase in the number and arrangement of the cells themselves, that is to say from quantitative-arrangement changes in the same entities, as with the photons of light or the atoms of the elements. Lobster and human nerve cells alike run on simple "binary based codes" operating in response to the frequency and duration of electrical signals.[4]

When it was first realized that the "current" running through the "nervous system" was electrical, it was assumed that it was simply continuous. But things turned out to be more complex than this. Near the ends of the long projections of each nerve cell – which under magnification look like thin, waving roots – lie bunches of molecules that convey the nerve cell's currents to the next cell and hence, are known as neurotransmitters. When the current flowing down a nerve cell strikes these molecule clusters, it creates an electrical turmoil that thrusts them into a gap (synapse) between the cell's branchings and those of the next nerve cell. There they are met by that cell's transmitters. It is through the intermingling of these molecular clusters that the current is transmitted from cell to cell. The emerging current can either "excite" the next cell's molecules to action or "suppress" them and block action. As with other cells, activity is based on negative and positive charges that cause some protein molecular patterns to stimulate activity and others to inhibit it. Nerve cells, in short, are a form of living matter and, like it, are activated by the clash of opposites. But they are a special form of living matter. Although they retain the basic aspects of the old, these aspects are not those of their viable essence but only provide its foundation. The situation is thus parallel to that between living matter in general and matter. Living matter, although containing particles and atoms, does not act in direct response to them but to the chain molecules containing them. Behind the functional complexity alike of matter, of living matter, and of living matter in its neurological form, however, lie comparatively simple mechanisms, based ultimately on the negative-positive oppositeness of the elementary particles.

If nerve cell chains end in the muscle fiber of a limb, their electric current can cause the limb to move. If they end in an internal organ – such as the heart – their current will motor the organ's functioning. If a certain area of a chicken's brain is directly stimulated by an electric probe, the chicken stands up as the current flows to its legs; another area and it will fluff its feathers. This will not occur if we simply prod its brain with a needle or stimulate it directly with heat, sound or light. Only with electrons. So, too, with the human brain. If a certain area is directly stimulated by an electric probe the subject sees; another area and the subject hears. Still other areas produce sensations of touch in various parts of the body. An ape and a man whose "visual" brain areas were alike electrically stimulated reached out as if trying to catch something, which the man identified as a butterfly. In short a flow of electrons through the nerve cells produces not just biological reaction but behavior, from a simple fluffing of feathers to reaching for a visioned object (in ape or human). And behavior is essentially a psychological process, produced, however, by a neurological one. We have crossed almost imperceptibly, from one level to another, from brain to mind.[5]

Brain activity does not, of course, normally arise from direct stimulation by electrons – from the neurologist's probe – but indirectly by the photons of light, molecular vibrations in air (sound waves), or the molecular formations that produce smell, taste or touch. These, the nerve cells convert into electrons by breaking down the molecules or combining the photons. Differences between sight, hearing, smell, taste and touch arise not from different kinds of internal stimulus; indeed, could not as all are reduced to electrons, but from the different paths in the brain followed by the electronic current. In short the (qualitative) difference between sight and taste arises from differences in unit arrangement. Further discriminations within each pattern of sight, hearing, etc. come from (quantitative) variations in duration and frequency in the electronic flow. In short, the brain – human or animal alike – acts in the same general bio-mechanical patterns as living matter does, and like it, is fired by the interpenetrations of opposite elements on various levels, from electrons and photons to negatively and positively charged atoms and atomic groupings, including molecular suppressors and exciters. Its special activity arises from the special shape of its constituent cells and nothing else.

Brain Research

How is it possible for the human brain thus constituted and weighing about three pounds to perform such diverse and complex functions, from bringing up children to writing books? The answer, it now appears, lies first in the extreme tinyness, immense number and rapid action of its operative units, from electrons to cells, and, second in the complexity of its organization, from divisions to subdivisions of subdivisions, all built up, little by little, in conjunction with bodily development in the long "trial and error" of evolution.

The figures are startling. The human brain contains some 20 billion nerve cells, more cells than there are stars in our galaxy. A dot, like the one at the end of this sentence, can contain 100 cells and the cell can contain trillions of atoms. Three pounds of such units represent an immense electrical complex. And this is only the beginning. Each nerve cell has 1,000 to 10,000 links with other nerve cells: "Microcircuits of electrons flowing from one neuron to another." It has been calculated (incredible though it may seem) that "the number of possible interconnections among these neurons is greater than the total number of atoms making up the entire universe." As Carl Sagan noted in 1977, "a modern computer able to process the information in the human brain would have to be about ten thousand times larger in volume than the human brain." When we realize the virtually infinite number of neurons and interconnections it holds within its patterned segments, the brain begins to become understandable – for the first time; and in purely materialist terms. It is, in fact, in itself the supreme example of the creation of qualitative change by quantitative and arrangement change. We think and feel and act – in short, are what we are – because of a combination of units that can neither think nor feel nor consciously act. As Lucretius long ago argued.[6]

The major segments of the brain are the same in humans as in other mammals. At the top of the spinal cord is the earliest evolutionary layer, shared with fish and amphibians, the brain stem or medulla, a kind of bulge at the end of the spine. The medulla controls a range of automatic activities from breathing to facial movements to the quick spotting of moving objects and, strangely, the stimulation of dreams.

Above the medulla is the cerebellum, whose two hemispheres control, among other things, balance and coordination. Above the cerebellum is the limbic (bordering) region which affects some major sets of emotions, especially those connected with sex and eating. Like the animals, we first react instantly and without conscious thinking to fear-producing and similar impulses hitting this region. The conscious thinking comes a fraction of a second later. Finally, above the limbic region is the cerebrum, with two hemispheres, the center in humans of memory, thought and imagination. All these major brain segments have their own identity and yet all interact. For instance, the quick spotting of a moving object via the medulla is almost instantaneously connected in humans with thought and conscious emotion (in the cerebrum); so, with balance and coordination (from the cerebellum), sexual reactions, fighting and eating and fear (all from the limbic region).

In recent decades, through surgical investigation and "brain imaging machines" more has been discovered about the brain's divisioning. For instance, when a neurosurgeon touched a small area of an epileptic patient's cerebrum with an electric probe, the patient cried out, "My mouth moved." Actually, as the surgeon observed, only one corner of the mouth moved. And even more minute areas of specialized function have been found. One set of nerve cells responds to the sight of a horizontal line, another to a vertical, still another, to a diagonal. Other cell groups respond to triangles, still others to rectangles. High-pitched and low-pitched tunes stimulate different sections of the auditory area in the cerebrum. "Brain-damaged patients ... may be able to write but unable to read, able to read numbers but not letters, able to name objects but not colors."

Using a "functional magnetic resonance imaging" machine, scientists have discovered, for instance, that the "region for spoken verbs is in the left frontal cortex, back of the left eyeball, deep down ... about the size of a pencil eraser." It has also been found that "there is a part of the brain that prepares for movement and a part that carries it out." The indication is that the human brain has been built up in the evolutionary process in extremely small specialized segments. So, too, with the animal brain. In a cat's brain some cells "fire" in response to a bar of light at one angle, others to one at a slightly different angle. "When a small area of the outer part of the hypothalamus is destroyed, cats will change from being easily handled to responding with 'blind rage' to all contact."[7]

Minute brain perceptions or reactions do not occur in isolation, either in animals or humans, but as parts of interactive complexes. Even pigeons, it has been shown, do not react to isolated stimuli but to stimuli perceived as parts of wholes: fish, other pigeons, water. In experiments with human brain imaging machines: "even simple tasks lit up several spots in different parts of the brain at once." When Wilder Penfield, the McGill University neurosurgeon, applied an electric probe to a certain area of a patient's brain, she remembered an old song from her childhood. Such memories, Penfield discovered, are parts of patterns:

> It is obvious that there is, beneath the electrode, a recording mechanism for memories of events. But the mechanism seems to have recorded much more than the simple event. When activated, it may reproduce the emotions which attended the original experience . . . It seems obvious that such duplicating recording patterns can only be performed in the cerebral cortex after there has been complete co-ordination or integration of all the nerve impulses that passed through both hemispheres—that is to say, all the nerve impulses that are associated with or result from the experience. It seems to be the integrated whole that is recorded.[8]

Just how anything is "recorded" was, until recently, a mystery. We can see how impulses can flow through nerve patterns but how can they be retained? What is the key – long sought – to memory? The answer, it now appears, lies not in some special, hidden quality of the nerve cells but in the fact that sensory impulses and thoughts etch physical changes into the brain, electronically creating trillions of new nerve cell connections – lasting pathways on a sub-microscopic level. When an impulse later hits these pathways, the original reaction that made them again appears. This process of building the immense complex of human memory is possible, of course, only because there is a virtually infinite number of potential neuron connections. It also puts in place an important new piece of evidence supporting the materialist view of mind as based on matter.[9]

It seems, from studies on premature babies, that the main connections between nerve cells crucial for human brain functioning arise not in "a smooth, continuous process extending throughout early fetal development," as had been expected, but "within a comparatively brief period of time," namely between the 28th and 32nd weeks. By birth the number of nerve cells has been set and does not change thereafter. What does change is the number and patterns of the links between them. At birth there are few connections, for instance, in the area of the brain that controls the legs; but by six years of age the area is a complex network of connections; and so, too, with language, sight and other areas. The changes in behavior patterns that we see in the development from embryo to baby and from baby to child reflect quantitative-arrangement changes in the brain that take place unevenly, sometimes in spurts. There are necessarily "nodal points" in such development as new (qualitative) changes arise. If a young monkey has a frosted contact lens placed for a time on alternate days on its left and right eyes, it will not later develop coordinated vision. The nerve cells needed for this do not develop after a certain age. So, too, with cross-eyed human infants. After two years, the condition usually cannot be reversed. The nodal point of change has been passed.

Behind quantity-to-quality changes and their nodal points there lies, as ever, the clash of opposites. Neuronic connections are not made either at random or in accordance with a master plan, but in response to negative-positive molecular groupings, which either allow a connection to be made or block it.[10]

Although the interaction between the brain and the rest of the body takes place primarily through the nerve network, there is also a secondary interaction of a very different kind, namely molecule groupings, known as hormones (Greek *hormaein,* to stimulate) which travel through the blood stream. One of these, thyroxin (an amino acid molecule plus an iodine molecule), which is made in the thyroid gland (in the neck), will in excess produce hyperactivity: its absence results in cretinism. The hormones, as Engels might have anticipated, act both individually and interactively. Letting loose the correct number of thyroxin molecules into the bloodstream is not the function of the thyroid alone but the thyroid in conjunction with two (opposite) hormones, one excitatory, the other depressant, released by the pituitary gland, located at the base of the brain. The pituitary gland seems to have general regulatory control over the other hormonal glands. It also produces the growth hormone. The epinephrine molecules released by the adrenal glands prepare the body for swift action – "flight or fight." Sexual behavior is rooted in the sex hormones, produced mainly in the testes and ovaries but also in the adrenals, which, in both sexes, produce male and female sex hormones. In short, psychological reactions in general have roots in the hormonal system, which is itself biological in nature. And the hormonal system interacts with the brain and the body's nerve system.[11]

The mechanisms of the brain, human and animal, have not only a long evolutionary history but, as Engels suggested, a kind of ur-history in one-celled creatures such as amoeba or bacteria. Such creatures can eat and digest plant and other biological matter, can approach or flee, and can mate with other cells (one cell injecting another with its nucleic acid). In approaching an object, they can respond to light, that is to say, to photons, encircling the object gradually in turns of about 30 degrees.

Their single cell, that is to say, has nerve cell potentiality. And there is a complete blending of biological and psychological process.

The Ladders to Mind

The next step up the evolutionary scale, one that apparently consumed many millions of years, is to connected blobs of one-celled creatures such as jellyfish whose "nerve-net" of "protoneurons" react as an undifferentiated mass to a stimulus anywhere on one of its constituent cells. Actual nerve cells first appear in such creatures as flatworms, which have a cord of such cells running down their bodies and forming a kind of knot at one end – an embryonic brain.[12]

When we ascend the evolutionary scale to fish – some 400 million years ago – we find a spinal cord within a backbone and a brain within a skull, the brain consisting of a medulla, limbic system, cerebellum and cerebrum (indicated also in the fossils of early fish). The fish cerebrum seems to be largely occupied with breaking down the molecules that cause smells and transmuting them into electrons; but in interaction with hormones it is capable of producing quite complex behavior:

> After establishing a territory and constructing a tunnel-shaped nest, the male stickleback "broadcasts" its sexual receptivity by swimming close to the water surface in a zigzag manner whenever a gravid female (laden with eggs) enters its territory.

These and other actions of fish are clearly non-conscious, for the fish cerebrum is far too simple to produce conscious thought. So, too, when we ascend the evolutionary scale to amphibia and reptiles. Lizards, crocodiles and other reptiles, including dinosaurs, are essentially biomachines reacting mechanically to stimuli in evolutionarily determined nerve cell patterns. Although psychological reactions and chemical reactions (in the nerve cells) are not the same thing, it is clear that there is no great gap between them. The process is similar in its essence to that of the stimulation of a chicken's brain with an electric probe producing physical reaction. Complexity in behavioral patterns reflects complexity of neural patterns.[13]

In birds, automatic behavior sometimes assumes considerable intricacy; for instance, in the mating dance of the albatross (the bird that plagued the Ancient Mariner).

> Male and female stand face to face with wings partly spread, surrounded by a large circle of shouting and clacking onlookers. The two performers raise their heads toward the sky, then duck to the ground only to rise again

touching beak to beak. The beaks are ducked first under the left wing and then the right, and heads are reared once more to the sky—dancers matching step for step. As the rhythm increases the clacking of the outer circle reaches a crescendo—and if the dancing male should tire or falter, another on the sidelines slips in and takes up the dance.[14]

Darwin is said to have remarked that such ritual is as complex as a symphony. Certainly, if we observed it in humans we would assume that it was a conscious process, that the dancers had consciously learned their parts and knew they were dancing. But, again, the lack of cerebral development in the bird shows that even so elaborate a ritual as this of the albatross is biomechanical. As such it illustrates the extraordinarily complex patterns of behavior that a brain operating on what must be an entirely or almost entirely non-conscious level can achieve. The birds may somehow sense that they are birds but they cannot "know" it or know that they are dancing. Their brains are transforming the particles, atoms and molecules of the external world into electrons and moving them through a maze of inherited neuronic patterns where they mingle with the molecular spur of the hormones. These patterns are inherited to lay a base for similar actions in succeeding generations. This base is then supplemented by imitation. Even at this level – which includes the mechanical mathematical calculations of bees – we can begin to see the conscious mind, as it were, in early embryo. We might note also that as birds have been in existence for 150 million years, such behavior must also have been going on for 150 million years, almost all of it without benefit of human observers.

When we come to mammals which, like birds, evolved from reptiles about 150 million years ago, we take a giant step forward. Mammals' brains are ten to one hundred times as large as those of reptiles of comparable size, and the increase is primarily in the cerebrum, which becomes – for the first time in evolutionary history – basic to behavior. If fishes' or frogs' cerebrums are cut out, they can still swim or jump much as before; but if a dog's cerebrum is removed, the animal is helpless, unable even to stand. Mammals, then, represent a major advance, the high points of which prior to humans were the apes – who evolved from monkeys – and the pre-human "hominids" such as "Lucy."[15]

Although mammals still perform complex mechanical tasks, such as the lodge-building of beavers, they have elementary reasoning powers and superior memory.

According to laboratory tests, fish can remember the location of food for no longer than ten seconds, but cats can remember for half an hour and monkeys for a day. The elementary reasoning that we can observe daily in dogs, or that Engels noted in the fox, can reach quite a high level in monkeys – as the story of Imo, the "genius" monkey observed by Japanese biologists, illustrates. To supplement the diet of a group of wild monkeys, the biologists threw wheat on a beach near the jungle. When the monkeys scooped it up, the sand made it virtually uneatable. Imo came to the rescue, throwing the sand-immersed wheat into the water. The sand sank but the wheat floated – and was scooped up and eaten.[16]

Apes can solve quite complex problems and, aided by sign language and other devices, can express associative ideas. In a classic experiment a banana was hung out of reach, from the top of a chimpanzee's cage, and three large boxes of increasing size were placed at random in the cage. The ape in time solved the problem by placing the boxes on top of each other in correct order of size and then mounting the crude stairway thus constructed to reach the banana. In another experiment, an ape was shown spots in a paddock where food was buried and a spot where a rubber snake was buried. Other apes were then allowed into the paddock. Within seconds the first ape had conveyed the information to its fellows by almost imperceptible gestures and eye movements. They went straight to the food spots and dug up the food. When they came to the spot where the rubber snake was buried they started back as though in fear. No doubt the early hominids had a similarly complex system of communication by gesture which we, of course, still retain to some degree in body language.[17]

Their natural signing method of communication can be further cultivated in apes by laboratory training:

> On seeing for the first time a duck land quacking in a pond, Washoe gestured "waterbird," which is the same phrase used in English and other languages, but which Washoe invented for the occasion. Having never seen a spherical fruit other than an apple, but knowing the signs for the principal colors, Lana, upon spying a technician eating an orange, signed "orange apple." After tasting a watermelon, Lucy described it as "candy drink" or "drink fruit."

Kanzi, a "pygmy" chimpanzee, learned to type requests for "ice to chew on, or for ice water or ice Coke" or for a favorite TV video. When he wanted the movie *Quest for Fire,* about primitive early humans, he punched "the symbols for 'campfire' and 'TV'." We should note, too, that apes have a sense of self-awareness. Unlike other animals, they can recognize themselves in a mirror. And like Kanzi, they can relate to TV – as dogs or cats cannot.[18]

The limitations of ape mentality have also been demonstrated. An ape learned to take a ladle and dip it into a barrel of water to extinguish a fire placed between it and some food. It was then taught how to paddle a raft in a pool. Next the ladle, the food and intervening fire were placed on the raft with the ape. The raft was then pushed from shore. Instead of dipping the ladle into the water of the pool, the ape paddled to shore, ran with the ladle to the water barrel, rushed back to the raft and extinguished the fire. The ape was unable to generalize (water), as a child soon learns to do.[19]

Apes also possess a rudimentary esthetic sense. If they are supplied with paint and brushes, they often work so obsessively that – like Shelley – they forget to eat. One ape, called Congo, produced 400 paintings.

> Although most of the efforts consisted of scribbling, the patterns were far from random. Lines and smudges were spread over a blank page outward from a centrally located figure.... Congo's patterns progressed along approximately the same developmental path as those of very young human children, yielding fan-shaped diagrams and even complete circles. Other chimpanzees drew crosses.[20]

Unfortunately, no subhumans are still extant (although it may be possible to create them), so we cannot directly compare their brains and behavior with those of humans. However, the apes' behavior must be generally similar to that of such subhumans as "Lucy" of some 4 million years ago, with a brain but little larger than those of chimpanzees. Ape behavior clearly hovers on the verge of consciousness; yet we hear no religious paeans about the "miracle" of the ape "mind."

Clearly there is a considerable gap between the automatic behavior of fish and birds and the quasi-conscious behavior of apes. Nevertheless the one must have grown out of the other; and we can observe an intermediate state in the non-simian mammals such as cats or beavers. It is apparent also that the differences must depend basically on increase in brain, especially cerebral, size relative to body weight.

When we move up the evolutionary scale to the "homo erectus" people of 1.7 million years ago, with brain sizes but slightly below the human, we must assume a virtually full potential for conscious thought and emotion. And this is borne out by the evidence that they used fire and engaged in the collective hunting of large game (which may indicate some use of language). Consciousness, then, as the observations of living creatures and the fossil record alike indicate, is something that grew in interaction with the body as part of the general evolutionary process, doubtless at times in relative spurts and smaller qualitative changes, until major qualitative changes took place.

This, too, we can observe in ourselves. It has become a kind of truism that "Man" has "consciousness." But fetuses and infants do not have consciousness, and very young children have but a kind of half-consciousness. When an infant smiles or laughs or cries or plays, it does not know it is doing so or that it is a human creature. It is simply responding in a non-conscious way to stimuli much as a cat or dog does. We read things into infant behavior on the analogy of our own behavior. When an infant is said to first "recognize" its mother, this is not a conscious recognition but the same kind of biomechanical recognition that, say, seagulls – as Tinbergen observed – have for each other. All of us, in short, begin as non-conscious beings, acting mechanically. We become conscious in early childhood, reflecting, not exactly but in wavering fashion, our brain's evolutionary history from the unconscious mating rituals of fish and birds to the quasi-conscious reasoning of the apes, and from the earliest hominids to *homo erectus.*

Chapter Six

BRAIN, BODY AND SEX

"Great social changes," Marx wrote to his friend Dr. Ludwig Kugelman in 1868, "are impossible without the feminine ferment." Historical advance, then, depends on both "feminine" and (implied) "masculine" social "ferment." And, Marx's wording indicates, he believed that there are some differences between the two. What these are and what the bio-social elements behind them are has begun to be unravelled in recent decades. Unfortunately some of these findings have been interpreted with an anti-feminist bias and this has led some feminists and proclaimed "Marxists" to a virtual denial of psycho-biological factors and a simplistic emphasis on social factors as the only relevant shaping force. It is time to redress the balance, to perceive interactions and not encapsulated units.[1]

Biological Roots

Although sex is determined by the genetic material in the x and y chromosomes, differences in behavior between the sexes in both humans and animals – our evolutionary ancestors – are directly influenced by the sex hormones. Experiments show that these hormones have deep shaping impact on body and brain. Female rats injected at the age of a few days with the male sex hormone, testosterone, were later unable to ovulate or exhibit female mating behavior. Young male rhesus monkeys normally mount about equal numbers of males and females, but if their mothers are injected with testosterone before birth, they mount only females. If men are castrated and injected with estrogen and progesterone, both produced primarily by the ovaries, they assume something of female bodily contours. Male rats deprived in the womb of testosterone show feminine qualities, and female animal embryos injected with testosterone will change to male. There is also a growing body of evidence indicating that homosexuality in animals and humans has a genetic base connected with a hormonal base that becomes manifest at a certain point in fetal development. The genetic component for male homosexuality is part of the x

157

chromosome and is apparently transmitted in the maternal line. Behavior, rooted in it, is shaped by the sex hormones. As with the pituitary and other hormones, then, the sex hormones are powerful, underlying behavioral determinants. However, as testosterone and estrogen are also produced to some degree in both sexes by the adrenal glands (above the kidneys) and both are needed for the proper functioning of both sexes, it is apparent that there are not absolute differences but an interactive shading of elements. And in humans these elements are molded in various ways by a complex of social forces. We are not dealing with absolute determinants but, as Marx said of instincts, with "potentials."[2]

In discussing dialectics with a friend who had just been married, Engels wrote as follows:

> At the same time you can always make the thing clear to yourself by concrete examples; for instance, you, as a bridegroom, have a striking example of the inseparability of identity and difference in yourself and your bride. It is absolutely impossible to decide whether sexual love is pleasure in the identity in difference or difference in identity. Take away the difference (in this case sex) or the identity (the human nature of both) and what have you got left? I remember how much this very inseparability of identity and difference worried me at first, although we can never take a step without stumbling upon it.[3]

The "identity" arises, as Engels did not know, from the fact that almost all the shaping genetic material in any species is shared by both sexes and this is supplemented by the sharing of the sex hormones produced by the adrenals. The "difference," as he also did not know, is ultimately due to the sexual genetic material (nucleic acids) in the x and y chromosomes and directly to the greater quantity of female sex hormones present in the female and the greater quantity of the male sex hormone present in the male. As Engels implies, the identity-difference phenomenon is general in nature and society. In fact it is simply another illustration of the fact that opposite entities have a common area of interaction. The proteins and the amino acids, for instance, have mainly the same constituents, differing in only one element. And they interact in the cytoplasm of the cell to produce the new life form. Of the specific biological factors underlying sexual male-female identity and difference, little was known in Engels' day but he would certainly have welcomed the new knowledge as revealing a biological (materialist) basis for psychological phenomena and an interaction of opposite elements.[4]

Some present-day "Marxists" gloss over behavioral and other differences between human males and females and argue that they can be virtually obliterated by "social conditioning." In all animal societies, however, sex is a basic factor, so basic that, along with the general behavior common to both sexes in all species, it produces radically different areas of behavior in males and females right up the evolutionary ladder, from fish to mammals. In all we have male sex frenzy and female selectivity manifested in often elaborate mating rituals such as that of the albatross. Moreover these forces are so powerful that they effect the course of evolution itself. Darwin's theory of sexual selection as producing little by little, such male characteristics as uselessly elaborate antlers and overlong birds' tails to win recalcitrant females has been substantiated in recent years, as I have noted, by experimental evidence.

That sex is an equally basic point of cleavage in human society as in that of our evolutionary forebears is but to be expected. A range of evidence indicates that this is so. At birth, human and animal babies alike differ not only in their sexual organs, sex hormones, and secondary sexual characteristics but in every cell in their bodies. Females normally have two x chromosomes in each cell, males one x and one y. At puberty the biological differences become even more marked in response to the hormonal flood. Girls develop breasts and boys' penises grow larger, both acquire pubic hair, girls begin to menstruate, boys produce spermatozoa, girls' pelvises expand and fatty deposits occur on the hips, boys grow facial hair, a deeper voice, body hair, larger bones, firmer muscles. Men are on the average taller than women, have more muscle and proportionally larger hearts, lungs, arms, hands, feet and shoulders. On the other hand, it is estimated that 120-140 males are conceived for every 100 females, 106 males are born for every 100 females, more males are born dead; females live longer than males, the female immune system is superior to the male. Males and females are more likely to contract certain diseases; for instance, heart diseases in males.[5]

In view of the intimate interaction between brain, body and mind, such extensive biological differences could not but produce psychological potentials that also differ in certain ways. In recent decades experimental evidence on such differences has been accumulating. Girl infants seem more responsive to sounds than boy infants but boy infants seem more responsive to light and to objects in space. Boy infants babble (at the age of four months) about equally to the mother's face or to a toy; girl infants normally babble only to the mother. Girl babies on the average learn to talk earlier and better than boy babies. Girls learn to read more easily than boys and in general have better "verbal skills." "Remedial reading" classes contain mostly males: 3 to 1. On the other hand, boys at an early age seem, on the average, better at mathematics and at solving three dimensional "visual-spacial" problems including mazes. Boys are more aggressive than girls and engage more in rough-house play. Boys are clumsy in comparison to girls; girls can do finer work, requiring coordination of hand and eye. Girls have greater "skin sensitivity." There are more left-handed boys than girls. There are differences between male and female psychiatric disorders. Males are more disposed to hyperactivity (five to one), females to depression (two to one), males to autism (four to one), females to anorexia (between 10 and 20 to one). These various psychological differences are reflected in social actions. Violent crime – muggings, robbery, murder and rape – are almost entirely male phenomena. So too, with violence in war or fascistic regimes. Attendance at pornographic films or sex shows is almost exclusively male, as is sadism, reading pornographic works, and violent sex attacks on children. Males are routinely aroused simply by the sight of the female body; females primarily by a specific male. Men are more excited by nude photographs than are women.[6]

As we consider these psychological differences, it becomes apparent that they cannot be caused solely by social conditioning. Some of them appear in infancy; others are so gross (two to one, three to one, five to one, ten to one) that social factors alone cannot account for them, and some of them also appear in our evolutionary ancestors, the animals, from reptiles to apes. Others have been tied to hormonal factors. It is the male sex hormone that decreases longevity and reduces the male body's defenses against disease, that makes boys engage in rough-house play, and males to be particularly subject to autism and hyperactivity and less able than girls to do fine handwork. The psychological role of the female sex hormones is "less clear" because they are more complex than the male, but it is apparently their action that stimulates both sexual activity and evokes mothering in all its complex manifestations. The male sex hormone stimulates sexual activity in both males and females, sometimes excessively in males, bringing with it irrationality and violence. "A y chromosome," writes Candace Pert, a prominent brain-research scientist, "is a real cross to bear. It's a predisposition towards angry, violent, competitive, macho behavior." Behavioral differences between the sexes can be molded in various ways by social forces but, like all biologically rooted behavior, cannot be eliminated by them.

As with hormonal-biological differences, psychological ones are not absolutes but matters of degree. Some girls are good at mathematics, some boys poor at them, some boys can read as well as girls, some girls read badly. To say that girls are not as aggressive as boys does not mean that girls are not aggressive. Males can suffer from depression, females can become manic. In short there are, as ever, variables within basic patterns.[7]

In recent decades research has revealed that hormones can produce neurological change in animals. The first discovery came in 1976 when researchers found that the brain of the male canary had two small groupings of brain cells three to four times larger than in the female. These centers control the canary's song (only the male canary sings) and they grow in the embryo in response to the male sex hormone. In the same period it was found that in male rats the back of the right hemisphere of the cerebrum was slightly larger than in females, and in females the back of the left hemisphere was slightly larger than in males. It was then found that the limbic system's hypothalmus (which affects the mechanics of fleeing, eating, fighting and sex) is "significantly bigger in male rats than in female ones." It was also discovered that the nerve cells involved differed somewhat in structure in the two sexes. Differences were also found in the neurotransmitters and enzymes in male and female animal brains. All these changes apparently also arise directly from hormonal action in the developing embryo.[8]

In humans it has been discovered that the "bundle of fibers" that connect the cerebral hemispheres is "much larger" in females than in males, a difference detectable in the fetal stage. This difference, it is suggested, might "be related to women's superiority on some tests of verbal intelligence." It could conceivably also provide the basis for the psychological research finding that human males tend to think and act primarily in response to one or the other cerebral hemisphere, females with more blending of the two. In all these matters we are, of course, only at the threshold but they do indicate that Candace Pert is justified in arguing, basing herself partly on her own research, that in time "we will be able to figure out the chemical coding [in the brain] for the differences between the sexes."[9]

These various research findings not only give further support to the materialist view of living matter as the base for psychological phenomena but they show that opposite element interactions and quantity-quality change dominate in sexual as they do in other phenomena. Sex hormones, like all hormones, influence certain brain areas; the brain, in turn, acts on the hormones, releasing them, inhibiting them, regulating them – all in response to activating and repressive mechanisms whose oppositeness forms positive and negative atomic patterns. Hormones demonstrate how differences in quantity (sometimes very small) can bring about major qualitative change, behavioral as well as physiological. And the total picture shows that, once again, we have both opposite elements, each with its own distinctive essence, and areas in common.

Social Repressives

These findings also show that human males and females are both further apart and closer together than has been realized and that each sex has special areas of ability. When we consider this latter finding in the light of the Marxist view of society and history, it becomes apparent that the potential contributions of women to society have been suppressed to an even greater degree than has been realized. What affect has this had on society and culture, on social thought and philosophy? Indeed, on Marxism itself?

As we can tell from existing hunting-gathering and early-type farming societies, women for tens of thousands of years were equally active with men in social and cultural life. They were, however, largely shut out from political power. The hunter-gathering community was normally governed by a group of male elders. The advance from hunter-gathering to farming society (from which all else flowed) was, George Peter Murdoch and other anthropologists argue, primarily the result of women's inventiveness, for it was women (and children) who did the food collecting in hunting-gathering societies. But, if so, it resulted in a further loss of power by women. Farming produced surplus, surplus produced trade and trade produced warfare, which increased male political power. It also produced male potters and weavers and other professionals whose centralized production enabled them to largely take over traditional home industries run by women. As large landholders devoured smaller ones, farming society developed into feudalism, apparently first about 3,000 B.C. in Sumer and Egypt. In feudalism – an extensive exploitive society spawning classes and escalating war – male political power was further consolidated. Although upper-class women secured new social and economic rights with the class cleavage of society, for the first time in history elevating some women above most men, the mass of women were hurled into slavery or serfdom along with the mass of men, where they were doubly exploited, as workers and as women. In the long trail of exploitive class societies – slave-commercial, commercial-feudal and capitalist – that followed, women remained cut off from central economic and political power; and still are. Hence, the kings and parliaments, wars and oppressions of "history," although presented as somehow neutrally human, have been male dominated. So, too, today. Today, also, the new nuclear and environmental threats to the life of humanity arise from ruling class males, not from "us."

In recent years feminist social scientists have been emphasising the special qualities of women. Women, they find, are "more interested in collective thinking" than men, "more interested in building support than antagonism." These and other special talents achieved new status in the Russian revolution and in building socialism. It was, in fact, a group of women, celebrating International Women's day, that by defying and cajoling the Czar's troops first got these troops to join the revolutionary forces. In the revolutionary war of the people of Vietnam against U.S. imperialist invasion, Communist women, fighting along with men, were often more successful than men in converting peasant groups and gaining support for the revolutionary army. In 1936, in early socialist USSR, women made their greatest single advance – a constitution that, uniquely, provided women equal economic and political rights with men.[10]

Feminist scholars also contend that women tend more to envision things (dialectically) "as a functioning whole" than "to reduce them to their component parts." Research has also shown that women are better able to make interconnection between apparently unrelated phenomena (which contains something of the essence of artistic creation) than are men. According to Jo Durden-Smith and Diane deSimone in *Sex and the Brain,* the evidence indicates that men have, on the average, superior "visual-spacial skills" and think more aggressively than women but women have a "more balanced and discriminatory brain." Over-aggressive thinking is clearly a two-edged sword. It can both drive to discovery, and block discovery by initial, hostile rejection of the new. It also appears that men tend more to abstract, women more to concrete thinking.[11]

As we consider these differences it becomes apparent that they are all matters of degree and are complementary, and that every society would benefit from an unrestricted mingling of them. In considering these differences we must not lose sight of the fact that most thinking, feeling and creative endeavor is simply human, neither distinctively male nor female. What has been suppressed is not only women's unique abilities but their general human abilities. The fact that women's abilities have been thus repressed in exploitive societies through the centuries suggests that the (male) rulers of those societies would rather tolerate decreased efficiency than grant power to women. Such power, indeed, is potentially dangerous to these rulers. That, it could, for instance, be used to disrupt war plans, was uneasily realized by the slave-commercial society male rulers of Athens – as *Lysistrata* (ca. 400 B.C.), with its female revolt against war, demonstrates. It also seems apparent that in exploitive-class societies with their endless conflicts, wars and repressions, including the repression of women, the male's hormonally-based excessive aggression provides a biological base for social advantage; and that childbearing and nurturing – often continuous – in such societies places women at a disadvantage. In short, there is a confluence of social and biological forces at work in exploitive societies, the biological sometimes providing a base for social exploitation. For instance, that childbearing and nurturing do not of themselves prevent women from playing an important role in government at various levels has been demonstrated in socialist societies in the present century; socialist societies, moreover, forced below their potential by external capitalist-world pressure.

Almost all written social thought since the rise of exploitive class society in Sumer and Egypt until recent decades was not only upper class, but male. Along with the general social advances forced from later capitalism, mainly by the working class, came some feminist ones, including voting rights and access to publication, but the fact remains that while throughout the history of civilization upper-class men needed to understand the economic and political systems that they ran, women were generally excluded both from the systems and the knowledge, technical and theoretical, needed to run them. Upper-class men used idealist philosophy as an adjunct to religion to control opposition to their rule, both from rival classes and from women. At times mercantile interests – from early feudal India to pre-revolutionary France – needed materialist views in their struggle with the greatlandowning forces. In the intellectual ferments thus generated in feudal, slave-commercial and early capitalist society, women were on the whole prevented from playing a direct role, although – from Sappho to Lady Muraski – they were able to break free sometimes to write literary works. And lower-class women produced a rich heritage of folk art. Women, however, did play an indirect role in most fields, from politics to social thought, whose significance has, indeed, to use Sheila Rowbotham's telling phrase, been "Hidden from History."

Even class-exploitive societies with their discriminations against women are, nevertheless, societies of men and women living and working together. In family life and community living, women's views and ways of thought have been central and sustaining. Even though these were seldom directly recorded they contained a complex mass of knowledge and artistry, whose extent is grossly underestimated, passed on from generation to generation, particularly by mothers to daughters. That women have had social and psychological views that differed from those of men is apparent from at least the days of the Athenian drama with its powerful women rebels, from Lysistrata and Medea to Antigone and Clytemnestra. It became directly evident in the works of Mary Wollstonecraft and more obliquely of Germaine de Stael some 200 years ago. Both writers had immediate predecessors, some of whom wrote and published, and these too had predecessors. All were fighting a stultifying oppression and forged their views as they fought (sometimes physically, as in the British and French revolutions). As literature and history alike reveal, women also recognized the weaknesses and dangers inherent in some male thinking and did what they could to counteract them, particularly by influencing sons and husbands. These efforts were seldom recorded or even recognized, but their ubiquity is clear from both earlier fragmented and later fuller statements. What we have, then, from social thought to philosophy, is not just male but male and female views with, however, a female component well below its potential and a male component dominant in key areas.

In considering the differences between male and female outlooks and ways of thinking, we must avoid a current tendency to exaggerate them. Men and women think in fundamentally the same way. Both have consciousness – as no other animal does – and reason consciously, sifting evidence and coming to conclusions. The differences lie in tendencies that can be molded to various degrees, but not eliminated, by social forces. Both men and women can think aggressively, both can think abstractly or concretely. The artistic creative process is essentially human and not male or female; but male and female imprints exist within it. In some ways there are greater ranges within the sexes than between them. Nevertheless, differences exist in outlook between the sexes in some areas and these have to be taken into account in considering both philosophical and social thought.

There was a considerable feminist movement in Marx's day – some of it with working class roots – and both Marx and Engels were not only aware of its views but championed them. (Marx's daughter, Eleanor, was a Marxist, a Marxist feminist and a labor organizer.) Feminist thought is integrated into Marxist class perspectives in Engels' *Origin of the Family, Private Property and the State,* some of it perhaps coming from his radical Irish wife, Lizzie Burns and her sister, Mary. By Lenin's day a powerful world-wide feminist movement had arisen and Lenin became a leader of its proletarian wing, both before and after the October revolution.[12]

The Marxist-Leninist tradition of struggle for equal rights for women has become an integral part of communist action and thinking everywhere. However, in theoretical Marxist works in the USSR, thought was assumed to be non-sexual in nature and origin. So, too, in various Marxist circles today. That this does not signify, as seems to be implied, that it is somehow neutrally "human" is indicated by the exclusive use of "man" in relation to it and the omission of women's – direct or indirect – contributions to it. The neutral mask hid, and hid as, a male chauvinist face.

In the examination of most aspects of reality, for instance in the natural sciences, there is little difference between male and female views or ways of thinking. But in most fields, although there is a large area in common, there are also areas of difference. This is clear in such social sciences as anthropology or sociology. Women anthropologists see primitive and other societies differently in certain ways than do men anthropologists. Both are needed for a rounded picture of the society. As literature testifies, men and women regard various aspects of personal life differently. The love poetry of Edna Millay or Elizabeth Barrett differs in attitudes, emphasis and insights from that of Shakespeare or Ovid. This is not a difference that can be bridged. Nor should it be.

What, then of Marxist social science and materialist philosophy? As with any body of thought, it is the product of that general reasoning quality – integrated with experience – that is common to both sexes. Again, however, there are areas that are viewed somewhat differently by men and women. Women see some aspects of society, humanity and nature in a different perspective than do men. Although some of this is apparent in some women Marxists, it has not been recognized for what it is or properly developed. This does not, of course, mean that there is a male Marxism and a female Marxism. But it does mean that Marxism, social and philosophical, has to become, consciously, the joint product of men and women. Otherwise certain aspects of society, humanity and nature will be inadequately revealed.

It seems clear also that Marxist-Leninist thinking about the role of the sexes in society has to acquire greater scope. It marks a qualitative advance over the essentially upperclass orientations of bourgeois feminism, particularly in perceiving the effects of class exploitation and oppression, but so far it has neglected the bio-psychological roots of these social factors. It generally presumes that we all begin life with a clean psychological slate on which virtually anything can be written. It fails to explore the basic area of the interplay of bio-psychological and social forces and thus lacks depth in some areas.

Chapter Seven

THE IMPRINT OF EVOLUTION

Although for some centuries "naturalists" have observed animals in the wild, it is only in recent decades that such observations have been scientifically controlled. Many of them, fortunately, have also been recorded in motion pictures. As some of the "sociobiological" views related to this research have been used for racist, sexist and imperialist purposes, many Marxists have not only rejected them but have denied the validity of the research. It is, however, becoming clear that the research itself is generally sound. When studies of animal societies are properly interpreted, they throw new light on the nature of people and further shatter abstractionist – ultimately religious – concepts of "man" and "mind." They show in some detail how we were shaped psychologically (as well as biologically) by the forces of natural evolution, as Engels long ago contended. They provide an evolutionary perspective for the findings of brain and psychological research, indeed for psychology in general, and they give new depth to the materialist view of the interrelation of people, nature and society.

Ape Society

The animal-society studies that have most direct relevance for human society are those on the African great apes, the gorillas and chimpanzees, which are, along with the orangutans, our closest living relatives. Two of these studies, that of Jane Goodall on chimpanzees and that of Dian Fossey on gorillas were initiated by Louis Leakey in the belief that they would throw light on early sub-human society. Leakey's belief has been substantiated by these studies and by one, also on gorillas, by George B. Schaller. When we read these books or see films on ape and other animal societies, we are not, as some commentators imply, simply looking at "clever animals" but at the evolutionary embryo from which we grew. Recent research indicates that our first upright ancestors evolved from an ape species between four and five million years ago. And it has been shown that our genetic material differs from that of the apes by only one percent.

Although our first direct hominid ancestors – the "Lucy" sub-people – have vanished, we can get a general picture of them and their life by comparing, say, Jane Goodall's films of chimpanzee society with films of hunting-gathering society; for instance, that of the Kung people of the Kalihari Desert in Africa or the Yanomani Indians of the Amazon rain forest (filmed by the Canadian broadcasting system). If we concentrate in both on basic behavioral patterns, we can begin to see how we became what we are.

Goodall's and other animal-society research not only shows that evolution determines our underlying psychological patterns, it is beginning to show it in detail. This is, admittedly, hard to grasp at first or to apply in our daily thinking. A few years ago I gave a lecture at the New School for Social Research in New York City on Marxism and Sociobiology. In it I noted that the games played by young apes – such as "follow the leader" – are similar to those of children today. This young ape behavior, I argued, showed why children all over the world act in essentially the same ways. A member of the audience protested: children in different cultures act very differently. This is true, of course, but only in regard to surface specifics. And I was not discussing these, but basic patterns which are very much the same everywhere and lead to the questions: Why are human children what they are? Which is simply part of the larger question: Why are *we* what we are? The key is to be found not in philosophical speculation but in ape and other animal society. The animal society findings of such observors as Goodall, Fossey and Schaller provide a unique window into our remote ("hominid") past. After looking through it we can no longer discuss "human nature" as a compound of "moral" and "intellectual" abstractions, either God-given or of unknown origin. We begin to see how we became what we are. The study of such works, in short, has profound, if generally unrecognized, philosophical import. They open a window on the remote past.

It also has basic social import, for what Schaller, Fossey and Goodall establish is that the apes have a society that resembles human society in some basic outlines. When our earliest sub-human ancestors emerged, some four million years ago, they inherited not only psychological characteristics but a societal structure of family life and work patterns – centered around food-gathering – from their ape forebears.

Schaller studied the gorillas inhabiting the forested mountains of central Africa (Zaire, Uganda, Rwanda). From early 1959 to the fall of 1960, he lived with gorilla groups, often sleeping near them at night, and made detailed notes on his observations. Contrary to the usual view, he found them on the whole to be gentle creatures who, once they were used to his presence, mostly ignored him. The gorillas lived in family groupings, the largest of 27 members (seven adult males, nine females, five juveniles and six infants). The groups moved through the forest eating vegetation. They made crude nests for bedding down at night. Each was led by an older male, a "silverback." The groups displayed hostility when they met. Schaller described an encounter between groups, one of them, which he knew well, led by a large male he called The Climber:

> Group VII met group XI on April 18. The Climber sat hunched over, staring at the ground in front of him, seemingly in deep thought, with his group clustered tightly around him. The silverbacked male of group XI squatted only twenty feet away by a tree, with the fifteen members of his group scattered through the underbrush behind him. This male was excited, and he put on a spectacular show: he hooted softly and with increasing tempo until the sound slurred into a harsh growl; he beat his chest, wheeled about, lumbered up a log, and with a forward lunge jumped down to land with a crash. As a finale he gave the ground a hollow thump with the palm of his hand. The Climber walked rapidly toward the other male, and the two stared into each other's eyes, their faces a foot apart. . . . The other members of the group paid little attention to the leaders; they acted as if they had no interest in the outcome of the struggle. . . . The Climber sauntered up to his opponent a final time and stared at him in a half-hearted way before wheeling around and walking rapidly away, followed by his group.

Such scenes give a sense of watching human actions in social embryo, here confrontational males challenging and backing off while others, although actually involved, feign indifference.[1]

The number and nature of such parallels between ape and human behavior, in all human societies, are so considerable that it becomes apparent they cannot be coincidental but indicate that our brains contain behavior-controlling neuronal patterns inherited from the apes. If we pursue the matter further and investigate other animal societies, we find that many of these patterns are shared by mammals in general and some of them also appear in other, still earlier, species. Our acquisition of speech and consciousness has, of course, altered specific patterns but the foundations remain. In short our basic forms of behavior are rooted in early ape society and more remotely in still earlier animal societies. Indeed, there is nowhere else they could come from.

This, as we shall see, does not mean that we are predestined to follow these exact patterns of behavior. Clearly our acquisition of speech and consciousness has altered specific patterns in various ways. And social conditioning, blending with these, has brought about still further changes. Nevertheless the roots remain and shape general directions in behavior.

Although there is cooperation within the groups, there is also, Schaller found, an order of dominance, partly based on age and sex but also on personality differences. Adult females are generally subordinate to adult males, younger males to older males, juvenile males and females to adult females, some adult males to other adult males, some adult females to other adult females. The leader of the whole group has unique eminence. A male lower in the dominance order will stand aside to let a male higher in the order pass on a narrow trail and will yield a seat on a log to him. Submission is shown in a typical posture:

> When by chance a game became too exuberant and rough, an infant showed this by crouching down submissively, arms and legs tucked under, presenting only its broad back to the opponent. An adult female receiving the worst of it in a quarrel with another female may employ the same submissive posture, and the other animal always respects the gesture and refrains from attacking further.[2]

Juveniles and infants, Schaller found, love to play games:

> Wrestling is a favorite pastime, and usually the arms and legs flail like windmills as the young roll over and over. Another frequent game is follow-the-leader, with the route going up trees, across fallen logs, down lianas, and perhaps across the belly of a dozing female. King-of-the-mountain is played on stumps and in bushes. As one youngster tries to storm the vantage point, the defender kicks the attacker in the face, steps on his hands, and pushes him down. Anything seems to be fair. Yet no one was ever hurt in such games, nor did they ever end in a quarrel.[3]

One long game-like sequence involved Max, "a rambunctious six-month-old infant who could never sit still" and an "old female" who "lacked an infant of her own."

> One sunny morning, as the female slept on her belly, Max climbed up on her back, walked forward and stood on her head, slid off, and then, as she rolled over, climbed upon her abdomen. She grasped Max and held him against her belly with one hand, and he struggled and squirmed and tried to free himself. His mouth was partially open, the corners pulled far back into a smile. The female loosened her hand, and Max grabbed it and gnawed at her fingers. She then toyed with him, touching him here and there as he attempted to catch her elusive hand. Finally Max lay on his

back on her belly, waving his arms and legs with wild abandon, and the old female watched the uninhibited youngster with obvious enjoyment. Suddenly Max sat up and, with arms thrown over his head, dove backward into the weeds.

Moritz, a seven-month-old infant, arrived on the scene. He walked up to the old female and pulled her leg. Max rushed from the undergrowth, tumbled over Moritz, and yanked his hair.[4]

The main care of the infants is done by the mothers, for instance "grooming":

When the youngster is small, the mother usually lays it in her lap or drapes it over one arm, then carefully grooms its pelage by parting the hairs. At such times her lips are pursed and her eyes watch her active fingers from a distance of six inches or less as if she were terribly shortsighted. She pays special attention to the cleanliness of the rectal region, which in youngsters is marked by a tuft of white hair like a fluff of cotton. The infants do not at all enjoy being turned upside down to have their anus inspected. They wiggle and squirm and kick, but the mother firmly persists with utter calmness.

The males are generally affectionate towards the infants:

The female leaned heavily against the side of the male. Her hairy arm almost obscured her spidery offspring, whose hairless arms and legs waved about in unoriented fashion. The male leaned over and with one hand fondled the infant.[5]

So far in psychology texts, human behavior has generally been treated as though it was without evolutionary roots – Freud, with his mechanistic biological orientation, was something of an exception – thus implicitly leaving the door open for a religious overview or a skeptical rejection of a materialist explanation for human origins. But Schaller's pioneering study makes it apparent that our behavior has evolutionary roots.

Almost twenty years after Schaller's book was published came *Gorillas in the Mist* by Dian Fossey, who lived for some thirteen years among the African mountain gorillas and apparently died in their defense. She found that the gorillas (to some of whom she gave names) not only lived in groups but that each group had a specific feeding territory. She noted in the first group she observed that two silverbacks "maintained protective positions around the females and young" and gave out "alarm" calls at any sign of danger. She found evidence of severe fighting, especially among the males – mostly apparently for the possession of females – and of infanticide by the males. She found more aggression among females than Schaller had, including one severe beating. And she found that dominance within the group was not absolute:

> On one rainy-day-resting period both Effie and Pantsy built large bathtub nests out of bushy *Hypericum* branches and settled down as comfortably as possible during the downpour. Beethoven [a dominant male] had made himself only a slipshod nest, into which he settled resignedly as the rain increased. About a half-hour passed before he began appraising the more comfortable positions of Effie and Pantsy. Abruptly he stood up, strutted to Effie's side, and stared down accusingly at her. Effie shifted positions and pretended to ignore him. With somewhat of a miffed expression Beethoven then strutted some twenty feet over to Pantsy and again stood in a stilted, intimidating position nearly on top of the young mother, leaving little doubt as to the nature of his intent.
>
> I fully expected Pantsy to ooze submissively out of the far side of her nest in obedience to Beethoven's postural command. Instead, she made it quite clear that he wasn't going to pull rank on her, looked directly at him, and harshly pig-grunted.[6]

Gorillas, Dian Fossey found, have a considerable "sense of curiosity," as when she was attempting rather awkwardly to climb a tree and a group came to observe her efforts, "sitting like front-row spectators at a sideshow." What, we can only wonder, was their mental state? They cannot have "known" that she was a human or what a human was but they must somehow dimly have recognized that she was a creature somewhat like them but unable to do the things that they could. There seems to be not only a kind of ur-curiosity here but an ur-amusement as well, perhaps reflecting a state of quasi-consciousness.

Fossey also noted, as had Schaller, that gorillas sometimes feigned indifference to events; and that staring was considered a hostile act (as it is in human society). Schaller had observed one or two copulation scenes. Fossey also saw male masturbation, "unisexual mountings," and sexual "advances" with the female taking the lead:

> Pantsy lay down beside the young male and, with the dorsal surface of her right hand, stroked his back and head. Icarus responded by reaching out and patting her arm hair gently, while wearing an interested facial expression. Eventually, he sat up and stared into Pantsy's eyes, his forehead furrowed questioningly and his lips contorted into a nondescript smile. Quivering, he propelled her rump toward himself and covered her body in a close embrace. The two exchanged prolonged sighs and soft,

humming belch vocalizations, apparently unconscious of the presence of myself, Effie, and her curious daughters.

Although the males helped with the upbringing of the young and were usually affectionate, the main nurturing was done by the females. Fossey, like Schaller, found that each gorilla had its own distinct personality; and that there were good and bad mothers: "Flossie was very casual in the handling, grooming, and support of both her infants, whereas Old Goat was an exemplary parent." The female juveniles, but not the males, early developed an interest in the infants and were sometimes allowed by the mothers to "cuddle, carry or groom" them. Apparently, then, we inherited from the apes not only general behavioral patterns but a tendency to individual personality traits (here underlying tendencies to careful or careless mothering and an early female special interest in offspring).[7]

Among the most extensive studies of African apes are those made by Jane Goodall in Tanzania on chimpanzees. That chimpanzee behavior is essentially, although not exactly, the same as that of gorillas is evident from her first observations on arriving at the area:

> While many details of their social behavior were hidden from me by the foliage, I did get occasional fascinating glimpses. I saw one female, newly arrived in a group, hurry up to a big male and hold her hand toward him. Almost regally he reached out, clasped her hand in his, drew it toward him, and kissed it with his lips. I saw two adult males embrace each other in greeting. I saw youngsters having wild games through the treetops, chasing around after each other or jumping again and again, one after the other, from a branch to a springy bough below.[8]

It soon becomes apparent, however, that the chimpanzees' group and family structures are more flexible than those of the gorillas. Instead of a family group of a dominant male, his females and offspring, the chimpanzee family consists only of the females and their offspring, although males and females live together in general groups. The females mate with a number of males, and it is not known who is the father of any particular infant. Although males and females are in the same group, there is more interchange between groups than among gorillas and a combining of smaller groups into larger ones. As a result, in larger groups there is often not one dominant male but a dominant inner circle of males. Nevertheless, although the group structure appears to be looser, it is still of paramount importance both for protection and feeding rights. Leadership is a basic drive – as illustrated in the anecdote of "Mike," a chimpanzee subordinate to "Goliath," the group leader, and to others. Mike became group leader by charging the group not, as was usual, with large branches but with two kerosene cans he had stolen:

> After a short interval that low-pitched hooting began again, followed almost immediately by the appearance of the two rackety cans with Mike close behind them. Straight for the other males he charged, and once more they fled. This time, even before the group could reassemble, Mike set off again; but he made straight for Goliath and even he hastened out of his way like all the others. Then Mike stopped and sat, all his hair on end, breathing hard. His eyes glared ahead and his lower lip was hanging slightly down so that the pink inside showed brightly and gave him a wild appearance.
>
> Rodolf was the first of the males to approach Mike, uttering soft pant-grunts of submission, crouching low and pressing his lips to Mike's thigh. Next he began to groom Mike, and two other males approached, pant-grunting, and also began to groom him. Finally David Graybeard went over to Mike, laid one hand on his groin, and joined in the grooming. Only Goliath kept away.

In time, however, Goliath returned, assumed the submission crouch and groomed Mike. Goliath then lost the typical slow swaggering walk of the dominant male.[9]

Unlike gorillas, chimpanzees hunt and kill:

> Running the last few yards, through some thick bushes I glimpsed Rodolf standing upright as he swung the body of a juvenile baboon above him by one of its legs and slammed its head down onto some rocks. Whether or not that was the actual death blow I could not tell. Certainly the victim was dead as Rodolf, carrying it in one hand, set off rapidly up the slope . . . Other chimpanzees in the valley, attracted by the loud screaming and calling that typifies a hunt and kill, soon appeared in the tree, and a group of high-ranking males clustered around Rodolf, begging for a share of his kill. Often I have watched chimpanzees begging for meat, and usually a male who has a reasonably large portion permits at least some of the group to share with him. Rodolf, on the contrary, protected his kill jealously that day. When Mike stretched out his hand, palm upward in the begging gesture, Rodolf pushed it away. When Goliath reached a hand to his mouth, begging for the wad of meat and leaves Rodolf was chewing, Rodolf turned his back. When J.B. gingerly took hold of part of the carcass, Rodolf gave soft threat barks, raised his arm, and jerked the meat away.

Although in this case one chimpanzee was the sole killer, chimpanzees (like lions) sometimes do group hunting, some treeing a prey and preventing its escape from the ground while others climb the tree:

> A second later another adolescent male chimp climbed out of the thick vegetation surrounding the tree, rushed along the branch on which the last monkey was sitting, and grabbed it. Instantly several other chimps climbed up into the tree and, screaming and barking in excitement, tore their victim into several pieces.

Nor was aggression limited to the males:

> The chase went on until Flo forced the female ["stranger"] chimpanzee to the ground, caught up with her, slammed down on her with both fists, and then, stamping her feet and slapping the ground with her hands, chased her victim from the vicinity.

Male aggression can sweep into a frenzy that engulfs control and sensitivity:

> All at once a series of pant-hoots announced the arrival of more ["stranger"] chimpanzees, and there was instant commotion in the group. Flint pulled away from the game and hurried to jump onto Flo's back as she moved for safety halfway up a palm tree. I saw Mike with his hair on end beginning to hoot; I knew he was about to display. So did the other chimpanzees of his group—all were alert, prepared to dash out of the way or to join in the displaying. All, that is, save [the baby] Goblin. He seemed totally unconcerned and, incredibly, began to totter toward Mike. Mike began his charge, and as he passed Goblin seized him up as though he were a branch and dragged him along the ground.
>
> And then the normally fearful, cautious Melissa, frantic for her child, hurled herself at Mike. It was unprecedented behavior, and she got severely beaten up for her interference; but she did succeed in rescuing Goblin—the infant lay, pressed close to the ground and screaming, where the dominant male had dropped him. Even before Mike had ceased his attack on Melissa the old male Huxley had seized Goblin from the ground. I felt sure he too was going to display with the infant, but he remained quite still, holding the child and staring down at him almost, it seemed, in bewilderment. Then as Melissa, screaming and bleeding, escaped from Mike, Huxley set the infant on the ground. As his mother hurried up to him Goblin leaped into her arms and she rushed off into the undergrowth.[10]

Chimpanzees form friendships in which reciprocal favors are sometimes performed. When Mike became the group leader, "J.B.", his "buddy" received increased status. Then when Goliath, the previous leader, refused to give up bananas to J.B., J.B. summoned Mike. Goliath then gave J.B. the bananas. On another occasion, a male chimpanzee came to the aid of a "buddy" in a fight with baboons.[11]

One of the most extraordinary sights Jane Goodall witnessed was a kind of rain dance:

> The rain was torrential, and the sudden clap of thunder, right overhead, made me jump. As if this were a signal, one of the big males stood upright and as he swayed and swaggered rhythmically from foot to foot I could just hear the rising crescendo of his pant-hoots above the beating of the rain. Then he charged off, flat-out down the slope toward the trees he had just left. He ran some thirty yards, and then, swinging round the trunk of a small tree to break his headlong rush, leaped into the low branches and sat motionless.
>
> Almost at once two other males charged after him. One broke off a low branch from a tree as he ran and brandished it in the air before hurling it ahead of him. The other, as he reached the end of his run, stood upright and rhythmically swayed the branches of a tree back and forth before seizing a huge branch and dragging it farther down the slope. A fourth male, as he too charged, leaped into a tree and, almost without breaking his speed, tore off a large branch, leaped with it to the ground, and continued down the slope. As the last two males called and charged down, the one who had started the whole performance climbed from his tree and began plodding up the slope again. The others, who had also climbed into trees near the bottom of the slope, followed suit. When they reached the ridge, they started charging down all over again, one after the other, with equal vigor.[12]

One other extraordinary event was a violent battle between ape groups, partly captured on film. The male apes in the groups suddenly began attacking each other. This time there was no roar and bluff. The slaughter went on relentlessly for days. After witnessing this scene Jane Goodall is said to have revised her view of chimp society as basically orderly and decided that the orderliness masked an aggression simmering always beneath the surface but held in check by pecking order and other conventions. Dian Fossey witnessed no such violence among the gorillas but she found gorillas – mostly males – that had been killed in fights with other gorillas and she found fight scars on the bodies of gorillas who had died from natural causes. There seems little doubt that Goodall's view is correct. Not only in ape but in all animal society aggression is apparently ever present – particularly among the males – but normally controlled by instinctive group-preservation patterns.

From Ape to Human Society

The fossil record is yet sketchy but bone and tooth remains of early apes, evolved from monkeys, have been dated at about 20 million years ago. Hence, the kind of ape society that Schaller, Fossey and Goodall observed has a very long history. For some 20 million years apes have lived in social, family-like groups and established feeding territories, dominant male group leaders have had supreme status and have swaggered and glared and roared at each other, and sometimes fought and sometimes backed off, less dominant apes have given seats on logs to more dominant ones and assumed the submission crouch, young apes have wrestled and played king-of-the-mountain, mothers have groomed infants and fathers fondled them, conciliating apes have groomed each other, males and females have alike repelled "strangers" and avoided staring, "buddies" have reciprocated favors and some apes have acted with ritualistic rhythm to the excitement of storms.

That ape society is also a form of animal society very much the same as that of our first upright evolutionary ancestors, the "Lucy" subpeople of some four million years ago, is indicated by the close parallels between the two species in bodily shape and brain capacity; so close, in fact, that a "pigmy" chimp and an australopithecus have been placed together in a roughly composite drawing. Only a fraction of a percentage point must separate their genetic materials. The Lucy sub-people, too, must have hunted collectively, using primitive language and weapons, lived in family groups, with mothers grooming the young, the children playing lively games, mothers and fathers happily nurturing babies, males struggling for group leadership, subordinate males yielding to dominant ones, coherence within the group and hostility between groups, with male group leaders trying to stare each other down, while other group members pretended indifference, and, doubtless, at times savage fighting between groups. The brain and behavioral changes brought about by uprightness and plains living doubtless modified these patterns in various ways but in general they must have remained about the same. That these general patterns continued up the evolutionary line through "homo habilis" and "homo erectus" along with growing speech and dawning consciousness is indicated by hunting-gathering societies that have existed into the present and have been researched and sometimes recorded on film. In these societies – virtually untouched by civilization – people live in groups, with cooperation within the group and hostility between groups, individual dominance structures within the group, with a leader or leaders at the top and orders of dominance below, the group usually being run by a council of male elders. The hunting and killing is done by the males, child-raising done mostly by the females, with help from the males, who are affectionate towards the children. The children play follow-the-leader and king-of-the-mountain, wrestling and chasing each other. We find rites of greeting and submission and, as Darwin noted for "Man," the same outline facial expressions as among the apes for different moods and emotions – smiles, tension, fear, even indecision.[13]

Ronald M. and Catherine H. Berndt in their pioneer study of a food-gathering people, *The First Australians,* write: "People try to teach him [the baby] to talk. They laugh and play with him, petting and teasing him, saying the same words over and over again until he repeats them." Although the central attention here is on language and the attitudes are conscious, the basic reactions of the humans to the child are much the same as those of the adult apes. So, too, with the children themselves. The Lucy sub-people children of four million years ago must also have played on the plains much as did the young apes in the forests. So, too, children in the *homo erectus* societies of a million and more years ago in Africa and Asia, and in early human societies 100,000 years ago. The fact that children today all over the world play some of the same games that the young apes play indicates a genetic and social inheritance of many millions of years.[14]

In ape societies some aspects of behavior clearly come from teaching and imitation but it has become apparent in recent years that this "learning" takes place on the basis of what Nikolas Tinbergen has called a "template" of instinct – evolutionarily inherited behavioral patterns. That such patterns can be inherited is clear in the lower stages of animal evolution. Amphibians, reptiles and birds from the moment of birth are capable of complex behavior. A cuckoo, placed as an egg by its mother in the nest of another bird, will on hatching push the other bird's eggs out of the nest, using its back in a complex series of work motions. Turtles on hatching in underground nests get to the earth's surface by cooperative effort – one turtle alone could not do so – and then walk, sometimes quite a long distance, to the beach from which their mother came. These things, as Candace Pert might say, indicate a "wiring" in the brain's neurons, a wiring that is inherited generation by generation through particular formations in the nucleic acid chain molecules. Whatever learning is then imposed on such instinctive patterns is a secondary factor, for this behavior advances comparatively little beyond that present at birth.

When we come to mammals, with their brains, relative to body size, up to a hundred times larger than reptiles, the situation is more complex, but both observation and experiment indicate that the underlying instinctive patterns still form a template for behavior. Let us look, for example, at wolf society, of which there have been a number of studies. Wolves live in groups (packs) that are led by a "dominant breeding pair." The pack stakes out its feeding territory (by urinating on its boundaries), hunts collectively (some driving out the prey while others wait in ambush), and has pack unity but hostility towards other packs and "stranger" wolves. The pack has a "well-defined pecking order." In an encounter with a subordinate, a dominant wolf stands tall and struts about with head and tail held high and ears erect. The subordinate, however, will slouch, with head and tail lowered and ears bent back (as in dogs also). Sometimes the "tail is held between the legs," and sometimes the subordinate wolf "rolls over on its back, bearing its stomach in an apparent plea for mercy."[15]

When we learn that similar behavior patterns are shared not only by wolves and apes but by all mammals, from sea-lions to lions, from polar bears to elephants – all recorded on film – it becomes apparent that they can only be explained on the basis of evolutionarily inherited instincts. There are, as we might expect, variations in specific patterns. For instance, in the dominant "pair" among wolves; but there is an unmistakable overall series of patterns shared in part or whole by widely differing species; group coherence and cooperation, group feeding territories respected by other groups, a group leader or leaders, "bluff" fighting, as with gorillas, self-sacrifice for the group, alarm calls, reciprocal favors, dominance, submission (with a submission posture), begging (with a begging posture), greeting ceremonies, initial rejection of the stranger, grooming and reconciliation ceremonies, aversion to staring, feigned indifference, delayed reactions, play (partly mock fighting), rhythmic displays.

In addition to these general characteristics are those peculiar to each sex. These, too, vary in specifics but general patterns generally apply: an aggressive seeking for mates by the males, including sex "displays" and fighting, selective choosing of a mate by the female, the main nurturing of the young by the female.

The nature of these various behavioral patterns, both the general ones and those of each sex, indicates roots in the evolutionary struggle. The leader and dominance systems strengthen the group – family or community – and cut down on harmful inner conflicts. Self-sacrifice and submission, hostility to other groups and to the stranger, conflict displays without fighting, alarm calls, reciprocal favors, grooming and reconciliation help to preserve the chain of life as much as do the teeth of the lion or the swiftness of the antelope. Some of these patterns of behavior are found also in bird and other pre-mammal societies. Which patterns are discarded and which retained by different species must depend primarily, as with physical characteristics, on environmental continuity and change. We have thus a series of evolving behavior patterns for all life forms, of which those found in the more advanced mammals, particularly the apes, have the most direct significance for humans.

As some sociobiological studies seem to imply a direct psychological inheritance of these inter-species patterns – under the aegis of a kind of genetic deity – it might be well to reiterate the elementary fact that what is inherited is not psychological traits but neuron patterns in the brain, which form the biological basis for behavior patterns. Changes in the nucleic acid chain molecules that form brain cells are in themselves purely chance affairs. Like changes in the body cells, those in the brain cells that assist the survival of a species in a given environment survive; those that do not, perish, and any behavior patterns that they shape likewise survive or perish. Brain patterns themselves must, of course, have evolved in conjunction with changing behavior. Again we have the dialectical interconnection of two factors, both of which are necessary but one of which (the brain pattern) is basic and the other secondary.[16]

Genetic and Social Inheritance

That the findings on animal societies and behavior in the last few decades must apply in some ways to humans is apparent from the fact that we evolved from the animals, most directly from the apes, whose immediate ancestors were the monkeys, whose ancestors, in turn, extended to presimians and beyond. The question is not whether these findings apply but *how* they apply, how they are integrated with consciousness and social forces. This is not a question that Marx and Engels could properly explore, for the basic data were not then available. But they were aware of it and indicated general directions. Marx, as we saw, believed, as did other 19th century thinkers, that humans, like animals, had "instincts," which he regarded, however, not as absolute determinants of human behavior but as underlying "tendencies." Marx also believed that there was a "general" character for human beings that underlay socially produced characteristics:

> To know what is useful for a dog one must study dog nature . . . Applying this to man, he that would criticize all human acts, movements, relations, etc., by the principle of utility must first deal with human nature in

general, and then with human nature as modified in each historical
epoch.

This passage appeared in *Capital*, on which Marx finished working in
April 1867. As he was by then well acquainted with Darwin's views,
writing to Engels in 1860, that *The Origin of Species* (1859) "contains
the natural history groundwork for our own views," he can hardly have
believed that "human nature in general" was shaped by anything but
evolution.[17]

Engels argued that evolution provided "the basis" for the "pre-
history of the human mind" from "the lower organisms right up to the
thinking human brain." The "mental development of the human child"
he concluded, was a "repetition of the intellectual development" of
the apes. Clearly neither he nor Marx felt that such views ran counter
to their social theories. They presumably regarded the two things –
inherited behavioral patterns and social dynamics – as separate phenomena, interactive yet affected by different sets of "laws."[18]

Once we recognize the main inherited behavioral patterns in animal –
particularly ape – societies, we see evidence of their presence everywhere in human society. That people live in groups of various kinds,
family, ethnic, religious, economic (unions, businesses), political and
so on we either simply take for granted or consider a purely social
phenomenon. But as we seek out group origins and compare them to
animal groups it becomes apparent that there must also be a bio-
psychological factor at work. Remains of *homo habilis* camp sites of
some two million years ago indicate group living. Such groups, however,
cannot have been deliberately formed because these sub-people's
brains were not sufficiently developed to have achieved consciousness.
Their instincts must have led them, as with the apes and the "Lucy"
subpeople before them, towards group living, and the path was paved
by a continuing history of such formation from these subpeoples,
perhaps for two or more million years. When fully conscious humans
first evolved, apparently by 100,000 years ago and perhaps much
earlier, they may have thought that they were establishing or continuing tribal and other groups entirely as a conscious process, but the fact
must be that they, too, inherited social grouping from the remote past,
and were predisposed towards such formations by instincts of which
they were unaware. And so on into farming and then into more developed societies. The path was, in fact, unbroken. When the English feudal
barons united to push for the Magna Carta (1215), they doubtless
regarded this simply as a deliberate decision. And to some degree, of
course, it was. But beneath the decision lay two hidden forces: those of

bio-psychological tendencies and those of objective social forces. Their inherited tendencies drove them toward the formation of groups in general; the particular kind of group that they formed was determined by historical factors beyond their direct control, namely those of a feudal society with a growing commercial component demanding internal peace and economic unification.

When we examine groupings in modern capitalist society, we find some of the same general characteristics as in animal societies, including group leaders and rungs of dominance. Although these tendencies are clearly molded by socio-economic factors, that biologically-based survival drives underlie them is shown by the elemental nature of group loyalties, familial, tribal, ethnic, racial and national, all of which can explode into irrational violence. So, too, with self-sacrifice for the good of the group. A remnant of the alarm ("mobbing") call, with its swift rallying of the group, seems to lie behind phenomena from the lynch mob to a national frenzy for war. We find, also as in animal society, feigned indifference and delayed reaction to threatening events, both of which may have roots in the early animal instinct for immobility in the face of danger. We also find the begging posture, the slow swagger of the dominant male – as in highway patrol cops – and the bowing submission of the dominated. It has also been suggested by Jane Goodall that the bow from the hip and even the casual nod of recognition are vestigial remains of the animal submission crouch. The killing frenzy that we find in the apes – as in the chimpanzees' savagely tearing a monkey apart – can in exploitive-class societies be escalated on a monstrous scale, as the long, bloody history of war and massacre, rape and torture reveals. The human creature everywhere, particularly the male, clearly gets an orgasmic thrill from killing. The degree to which this instinct is developed, of course, depends on social factors but the instinct itself is always there. So, too, with greed. "Rodolf," the chimpanzee, after slamming the head of the young baboon against the rocks, greedily eats the carcass, defying the other chimpanzees "begging for a share of his kill." This instinct, too, can either be curbed or escalated by social forces.

Animals, from swans to gorillas, sense that leadership is vital to group survival. So, too, in human society, where it pervades more of our behavior than at first appears. There seems, indeed, to be, as some sociobiologists have contended, an urge in people to obey or even exalt a leader. This instinct, as history shows, can be channeled in either progressive or reactionary directions, lying indifferently behind a semi-psychotic worship of fascist demagogues or an exalted love for revolutionary leaders. Although such love arises directly from the accomplishments of such a leader, its electric vitality exposes deeper springs.

Leader exaltation also affects cultural judgments, as we tend to elevate certain thinkers or artists (poets, musicians, actors) to godlike status. We seem to have an urge to defend or support such persons absolutely. This applies to Marxism also. When we grasp the historical vistas and scientific depth of Marxism we feel an exaltation which, unless combined with judgment, can be channeled into virtual deification and thence into dogmatism. This tendency to cultural semi-deification also spreads beyond individuals to groups. And it can be perverted in exploitive-class societies where "the rich and famous" form a kind of hierarchy of the gods.

As we survey society, it becomes apparent that not only some but all our general behavioral patterns come from our animal, and especially our ape, ancestors. This includes everyday gestures, such as embracing, fondling after separation, holding hands, and kissing as well as the broader patterns we have just noted. It also includes the forms and even the techniques of parenting – as in systematic anus inspection by mother apes. And it includes auntlike behavior patterns – as in the relation of the childless older female ape to the rambunctious "Max."

This, however, is but part of the story. Our basic emotions are also rooted in the animals: love and hate, anger and defiance ("Pantsy" and "Beethoven"), exaltation and despair, grief for the dead and joy in children, love of play and love of the hunt (in various transmuted forms). And violence and greed, and self-sacrifice. Our patterns of thought lie there also, as it were, in embryo, as witness the apes, with their endless curiosity and their capacity for analysis and synthesis, from building a stairway of boxes in a cage to fashioning a twig as a tool for the capture of termites. And even our artistic urges are there, the joy in imitation, in painting, in the dance – as witnessed by Jane Goodall. All lie beneath the surface of consciousness and blend with its patterns. In short not just certain traits but the essence of our life wells up from our evolutionary past and nothing else. Examining this past we can see better both our potentials and our limitations and why we have both.

Sexual Behavior and Evolution

As we have seen, the sexual behavior of males and females in animal societies is directly determined by the sex hormones (acting through nerve cell patterns in the brain). The males, agitated by testosterone, drive frenziedly to copulation, a drive that everywhere, from birds to apes, leads to "display" and fighting. The females, responding to estrogen and progesterone, go through sexual cycles, and "choose" mates. When mating arrangements have been made – generally one male and one female – and offspring arrive, nurturing is mostly done by the female. There are variations to this pattern but it is the general one. It is easy to see that testosterone lies behind the male sex drive but it is not yet clear how the more complex female sex hormones govern the selection process, especially in animals with virtually no cerebral development, such as fish or amphibians. It seems impossible that the females of such species could directly grasp, as some sociobiologists seem to imply, which males will provide best for their offspring. In fact experimental evidence indicates that they are automatically responding to male physical characteristics, such as long combs and bright wattles in chickens, loudness of the song of the bullfrog, long tails in some male birds.[19]

The nerve cell patterns in the brain that the sex and other hormones operate through are, like all cell patterns, inherited from generation to generation and changing as species evolve. The mammalian patterns are more complex than the reptilian, ape patterns more complex than those of less developed mammals. And so with the human in relation to the ape. In the ape the patterns are linked with a quasi-consciousness and in humans with both consciousness and social forces. But basic instinct patterns are still clearly present, inherited along with, and intermingled with, hormonal and other biological differences. By and large human males display and compete and females actively select, although in some societies both processes are thwarted or distorted. It may also be that instinctual patterns for selecting mates lie behind such human female characteristics as sharper psychological insights and superior ability to recognize faces and voices. As Candace Pert noted: "Women . . . need to choose a mate who will stay around and take care of them and their offspring. So I'd expect to find a part of the female brain that is devoted to making that kind of choice." We find similar deliberate mate-choosing among human males. Nevertheless, the instinctual patterns that evolved in animal society supply only underlying directional drives for mate selection of which we are mostly unconscious.

In human as well as animal society, nurturing is on the whole mainly carried on by mothers, usually occupying a large part of their lives. The indication is that an instinctive base exists here also. This, however, is but the beginning of the story. All the basic psychological differences between the sexes that we have noted have evolved from our animal ancestors. Nor could it be otherwise. Like all human qualities, there is nowhere else for them to have come from. The evolutionary process, however, once again reminds us that we must not stress only differences. The adrenal glands ensure a mixture of sex hormones. In early evolutionary stages maleness and femaleness are sometimes reversible. In some fish, for instance, females can become males and males become females under conditions of extreme stress. This is not possible in higher species such as mammals. Nevertheless, male mammals, as we saw with wolves or great apes, can become less or more male under psychological pressure; again an example of the interaction of psychological and biological factors.

In humans, as Engels noted, the general "identity (the human nature of both)" is greater than the "difference (in this case sex)." The differences, as I have noted, are not absolutes but a matter of degree, based directly on a mixture of hormonal elements arising from the adrenal glands and created in the evolutionary process.[20]

Marxism and Sociobiology

In exploitive-class societies from the earliest feudalisms to monopoly-finance capitalism, exploitation and oppression have not only an economic but a bio-psychological base. The ruling classes, even though they are not fully aware of what they are doing, manipulate innate behavioral patterns. They use male aggressive and sex drives to herd young males into war. They encourage them to fight each other and not "the boss," economic or political. They use group-coherence, anti-outsider, self-sacrifice and pro-leader tendencies either to escalate the war process to a frenzy or to consolidate their rule – from Caesar to Hitler. They distort female biopsychological leanings to form a base for exploiting women economically and sexually. The same patterns also lie (indifferently) behind human progress, mass struggles and revolutions, past and present. Without such innate tendencies as leader-following, group (class) coherence and self-sacrifice there could have been no October Revolution and no struggle for socialism.

Artistic and cultural life is everywhere shaped, in its basic under-pinnings, by our evolutionary inheritance. The roots of our esthetic sensibilities are visible in the apes' paintings. The young apes' play, with its mock-fighting, reveals the patterns that underlie dramatic presentation just as animal ritual suggests a genetic base for human ritual (religious, military and so on). Such general fictional themes as a struggle for power (leadership) or love and conflict between the sexes clearly have an ultimate base in instinctive patterns. We have an innate "feel" for such things.

Our tendency to exalt a leader seemingly lies behind the urge to create and worship a god or gods – extended to religious leaders – and is a formative force behind religion in general. We can also see the roots of morality and moral judgments in the need for group coherence, with its accompanying reciprocal-favor behavior. Morality, that is to say, is neither god-given nor evolved by social needs alone but is rooted in the instinctive tendencies behind such needs, primarily the drive for group survival. We instinctively tend to condemn conduct that is contrary to the welfare of the group and support conduct that is beneficial to it. Our – often obsessive – sense of guilt may have roots in an instinct-created taboo against countering the interest of the group, family or otherwise. Moral judgments and the turbulent emotions often connected with them must arise from similar patterns. In short, animal society research has given us a new weapon to refute both the idealist view that the basis for morality resides in "the Divine" and the simplistic social-determinist argument that it has a purely social base.[21]

Even more broadly, we are provided with a materialist explanation for the foundations of human nature. We can no longer speak of "man" in the abstract and deduce "his" characteristics from general principles. It is now becoming apparent that humans have been built by evolution in all aspects of their being. The notion, direct or hidden, that the "mind" is a unique creation shaped by God (the Absolute, the One) has been further routed by the accumulating evidence of evolutionary psychological origins, evidence that supplements that on the interaction of body, brain, and mind. So, too, the idea that "the mind" is shaped in its essence by "social forces." As we perceive the detailed ways in which the brain originated in the evolutionary process and unite this with modern researches into its development from infancy on, it becomes evident that behind social molding lie inherited biopsychological patterns. We are not at birth simply blank tablets on which anything can be written but come with an elaborate under-pattern that directs the "wiring" of the brain as its neurons form their trillionfold interconnections. It is for this reason that we are able to feel joy and sorrow, loyalty and hatred, fear and hostility; in short, to be part of the stream of life. If there was nothing – as some "Marxists" really imply in spite of hedging disclaimers – but social molding on a brain virtually undirected at birth we would be, if anything at all, emotionless automatons. Although Marx and Engels were aware of these things only in a general way, they would certainly have accepted the new evidence as in line with their beliefs in a general "human nature" and human instinctive "tendencies" and put it in materialist perspective. Curiously enough much of the evidence was there all the time. It simply needed systematic observation of animal societies to bring it to light.[22]

Marx in his famed summary statement of 1859 depicted human society as a spiralling interacting sequence from socio-economic to cultural dynamics:

In the social production of their means of life, human beings enter into definite and necessary relations which are independent of their will—production relations which correspond to a definite stage of the development of their productive forces. The totality of these production relations constitutes the economic structure of society, the real basis upon which a legal and political superstructure arises and to which definite forms of social consciousness correspond. The mode of production of the material means of life determines, in general, the social, political and intellectual processes of life. It is not the consciousness of human beings that determines their existence, but, conversely, it is their social existence that determines their consciousness. At a certain stage of their development the material productive forces of society come into conflict with the existing production relationships, or, what is but a legal expression for the same thing, with the property relations within which they have hitherto moved. From forms of development of the productive forces, these relationships turn into their fetters. A period of social revolution then begins. With the change in the economic foundation, the whole gigantic superstructure is more or less rapidly transformed. In considering such transformations we must always distinguish between the material changes in the economic conditions of production, changes which can be determined with the precision of natural science, and the legal, political, religious, aesthetic, or philosophic, in short, ideological forms in which human beings become conscious of this conflict and fight it out to an issue.[23]

The nature of the "productive forces" – such as the windmills and plows of feudal society or the blast furnaces and tractors of capitalism – determines the nature of the "production relations": the greatlandowning system and peasant labor of feudalism, the capitalist ownership and industrial-worker labor of capitalism. These production relations in turn determine the general nature of the political and legal structures: the barons' councils and absolute monarchies of feudalism, the bourgeois democracies or political dictatorships of capitalism. They also ultimately shape some of the ways in which people think: "definite forms of social consciousness" – "legal, political, religious, aesthetic or philosophical . . . ideological forms." In class societies these "forms" are directly created in the maelstrom of class struggle.

As the economic and social foundations of society change, the class struggle changes and with it people's ideas on political, religious and other matters. People in general are not aware of the basic reasons for these changes because they are not aware of the function of the underlying socio-economic structure. The "ideological forms," we should note, do not comprise the whole of "social consciousness" but only a part of it. The rest of it consists of the ideas and emotions of personal, family and everyday life; and these are rooted more deeply in bio-psychological elements than in economic structures.

Clearly there is nothing in this view to contradict the existence of instinct or the evolution of the brain. There are simply two different areas involved; historical process and the nature of people, each with its own central dynamic. The areas, however, although distinct, have an area of overlap between them. Social life reflects the nature of people, a nature whose foundations were shaped by evolution. Inherited behavioral patterns play a basic part in family life and community life. The nature of people is, of course, reflected in the historical process. Even though the "production relations" provide the basic shaping force, they are created by people and hence bear a human stamp. So, too, does the class struggle, which no matter what its central socio-economic dynamic, is still a struggle of people.

The fact that people have innate behavioral patterns that are shaped in various ways by social forces does not, as some imply, contravene Marx's argument that the socio-economic processes constitute "the basis" for historical evolution. The change from feudalism to industrial capitalism in Europe that Marx here in part reflects (obliquely) was not directly shaped by human instincts but by socio-economic dynamics over which people had but indirect control. Yet without innate human behavioral patterns these dynamics would not have arisen in the first place. Imperialist war is obviously not, as some sociobiologists argue, caused by aggressive instincts but by the contradictions of capitalism. Nevertheless, unless imperialism had such instincts to work with, people could not be propelled into war. The two components, socio-economic and bio-psychological, interact, with the socio-economic playing the directly determinant role, the bio-psychological providing the raw material with which it works. Engels came close to seeing this when he wrote in *Feuerbach* of "the driving powers which – consciously or unconsciously, and indeed very often unconsciously – lie behind the motives of men who act in history." He would not have been surprised to learn – indeed perhaps assumed it – that these unconscious "driving powers" are rooted in instinct.[23]

It is apparent that although people are molded by social forces, they are not indiscriminately malleable; the molding has to take place within the framework of inherited behavioral patterns, particularly those of group and anti-group formations. The specific patterns of such formations shift and change but the general contours remain and weave their way through the spirals of history. If we perceive in their contours the bio-psychological roots of group phenomena, we can better combat those phenomena with reactionary thrust and build up those with progressive thrust. We can make use of this knowledge in organizing revolutionary movements or in building socialist societies. Marxists working in a capitalist or other exploitive society should be able to perceive movements based on group-loyalty patterns at an early stage and, realizing their elemental nature – as shown, for instance, in nationalistic frenzies – better assess how to develop or retard them.

Animal society research, then, for all the reactionary implications of some of the theories imposed upon it can, if used in a Marxist perspective, become a progressive force. Its findings, in enabling us to understand better what we are and how we came into being, give new depth to the struggle against idealism and superstition. They enable us to see more exactly how our evolutionary past determines our present and to perceive non-conscious directional patterns behind our thoughts and actions. In short, we need a Marxist sociobiology as an intellectual weapon to complement such already existing Marxist social sciences as economics, sociology, political science and anthropology.

When the findings of animal-society research are perverted for reactionary purposes, Marxists must, of course, expose the perversion. Yet these same findings, if viewed in a Marxist perspective, enable us to break with the reactionary mechanistic view of human nature as a blank tablet and perceive an underlying psychological oneness to animal and human life on our planet, much as recent genetic research has revealed a biological oneness. They uncover, exhilaratingly, the vital links between nature, people and society. Combining these findings with those of modern physics, biology, and brain research, we can begin to see the dynamic unfolding of the chain: from preatomic matter to atomic matter to – on our planet – living matter, to animal forms, to animal society, to subhumans, to humans and the human brain. Then, if we are Marxists, we can take the next step and see how human society grew out of these preceding stages and how it has since evolved socially.

Humans inherited hunting-gathering societies from prehumans, the whole going back some four million years. At some point hunting and gathering in various areas began to be supplemented and then super-ceded by farming. Farming produced surplus on a new scale and this in turn produced vastly increased trade. These developments led to feudal civilizations – as smaller farming units were devoured by larger ones. The large-scale farming of such early feudal states as Egypt and Sumer produced extensive commerce. Commerce stimulated industry (lumbering, quarrying, mining) and created a financial system (of banks with foreign offices). As commercial wealth increases more rapidly than agricultural wealth, capitalists in various areas at various times set up separate city-states – from Athens to Venice to Osaka – and later seized control of nations: Holland, Britain, the United States of America; and developed an industry based on coal, iron and steam power which required an industrial working class. Once again the tendency of smaller units to merge into larger ones, this time industrial and not agricultural units – asserted itself and capitalism grew into corporate capitalism, which led to wars and revolutions on a new scale, some of the revolutions producing a new form of civilization – socialism, in which the industrial working class was the dominant force. In brief, not only biological life and the human "thinking brain" have evolutionary histories which explain their existence but human society does also. We can perceive our remote past in our household pets and our psychological evolution in our children. We can see how human society emerged from sub-human society and the path it has traversed to the present. We can also see the forces, social and natural, that will determine the future. Some of these, unanticipated by Marx and Engels (or Lenin), present both new perspectives and terrible challenges.

Chapter Eight

LIFE, DEATH AND PURPOSE

Lenin paraphrased with approval the Latin poet Statius' line: "primus in orbe deos fecit timor." ("It was fear that first made gods in the world.") Although from the earliest human societies, with their chiefs and shamans and ruling male elders forming an embryonic class structure, religion has been an instrument of rule, it clearly could not have become so without fear, especially that engendered by death. In civilized societies the death theme has traditionally been central to religious doctrine and ceremonial. Christianity early adopted a language of symbolic innuendo that both reflected and heightened the human dread of death, a language that, blended with the solemnity of black robes, gloomy cathedrals, stained glass windows, somber music and crucifixes, created an atmosphere conducive to control by fear. The doctrines of sin and damnation and such rituals as transubstantiation center on death. Much religious art is sadistic and terrifying, from the hideous devils of Asian religions to the Christian torments of the damned or the execution scenes of Jesus and the martyrs; and even when such art depicts saintly or angelic bliss, the suggestion of death is not far behind. So, too, with such ostensibly positive themes – in words or music – as salvation or resurrection. Although death is not, of course, by any means the only concern of religion, it is clearly so central that without it there would be no religion.

On the other hand, materialists, dialectical or otherwise, have paid comparatively little attention to death, placing the emphasis on life and its roots in nature. But more balance in considering the blend of life and death is clearly needed to shape a more rounded materialist view and make materialism a more viable personal philosophy.

The Christian theological argument on death was creatively condensed by the poet Samuel Taylor Coleridge, who was also a clergyman, in his little poem, published in 1817, *Human Life: On the Denial of Immortality:*

> *If dead, we cease to be; if total gloom*
> *Swallow up life's brief flash for aye,*
> *we fare*

> *As summer-gusts, of sudden birth and doom,*
> *Whose sound and motion not alone declare,*
> *But are their whole of being! If the breath*
> *Be Life itself, and not its task and tent,*
> *If even a soul like Milton's can know death;*
> *O Man! thou vessel purposeless, unmeant,*
> *Yet drone-hive strange of phantom purposes!*

If when we die we cease to have any existence, we are like gusts of wind, nothing but "sound and motion." If such winds – material phenomena – are all there is and not the manifestation of ("declare") a spiritual presence (God), life is without purpose. This strikes home with special impact if we think of a great man such as Milton being annihilated. The "breath" and "tent," metaphors are essentially the same as that of the wind; "tasks" adds a moralistic note: if the phenomena of life are not just the outer (delusive) shell of existence but the whole of it, not just a burdensome "task" imposed by our material (evil) nature, but the essence of that nature, then, life is meaningless: "phantom purposes."

Coleridge goes further. If, he argues:

> *Thy laughter and thy tears*
> *Mean but themselves . . . Why rejoices*
> *Thy heart with hollow joy for hollow good?*

There can be no real "good" or "joy" in life except that arising from faith in divine purpose and immortality. Otherwise even the "tears" of grief are meaningless.

The British scientist, J.B.S. Haldane, in his pioneer study, *The Marxist Philosophy and the Sciences,* writes that although life and death "form part of a whole . . . we have no word with which to express the unity which they form except the word life." The word "life," that is to say, is used in two senses: to contrast life and death and to include life and death as part of a continuing life process. The semantic paradox indicates that people in general place the emphasis on life rather than on death. Coleridge, on the other hand, places it on death, actually isolating death from the total life process. We can see the segment of truth in his argument if we do the same and exclusively envisage the devastation of death in human society: the often long agonies of dying, the decay of the body after death, the untold billions of the dead and of those yet to die, the separation forever of husbands and wives, friends and lovers, parents and children, and the emptying of life this can bring to the survivors. Facing this true but truncated picture, which leaves out birth and growth, love and joy, triumph and struggle, Coleridge contends that life has no meaning except that supplied by delusion, especially the doctrine of immortality. He also implies that people should renounce sexual and other "hollow" "joys" and devote themselves to Spiritual abstractions. The "Garden of Love," William Blake – Coleridge's contemporary – protested was being transformed into a cemetery by such views:

> And I saw it was filled with graves,
> And Tomb-stones where flowers should be;
> And Priests in black gowns were walking their rounds.
> And binding with briars my joys and desires.[1]

Such a doctrine, Blake contends, is psychologically destructive and socially repressive. Moreover it discourages progressive political activism (as it did in Coleridge himself – as he turned from Utopian visionary to Tory apologist). The doctrine, including its central thesis of immortality, is for all its subtle eloquence not only false but reactionary, breeding both delusion and social apathy.

The classic arguments of idealist philosophy for immortality, although having roots in Hinduism, were first formulated by Plato about 400 B.C. in his dialogue, *Phaedo,* with, as usual, Socrates as the principle exponent. Socrates contended that the soul or, more exactly mind (the Greek philosophers did not distinguish between the two), is absolutely different from the body. The mind is akin to the unchangeable and indivisible "essence" of being, the body to matter. "Men or horses or garments" are "always in a state of change" but "essence or true existence, whether essence of equality, beauty or anything else" is "self-existent and unchanging." The body is a substance that can be divided, the mind is an indivisible "essence." Hence the mind can survive death but the body cannot. Moreover the mind must have existed before birth; otherwise we could never have acquired a concept of "the essence of equality" or of "beauty" for we have never seen them in life:

> But when did our souls acquire this knowledge? — not since we were born as men?
>
> Certainly not.
> And therefore, previously?
> Yes.
> Then Simmias, our souls must also have existed without bodies before they were in the form of man, and must have had intelligence.

Socrates then draws the implication that if the mind existed before birth it must continue to exist after death, when it will – shedding the "evil" body – pass once more into the realm of eternal spirituality with which it is akin:

> The truth rather, is that the soul which is pure at departing and draws after her no bodily taint, having never voluntarily during life had connection with the body, which she is ever avoiding, herself gathered into herself;—and making such abstraction her perpetual study—which means that she has been a true disciple of philosophy; and therefore has in fact been always engaged in the practice of dying. For is not philosophy the study of death?
>
> Certainly—
>
> That soul, I say, herself invisible, departs to the invisible world— the divine and immortal and rational: thither arriving, she is secure of bliss and is released from the error and folly of men, their fears and wild passions and all other human ills, and forever dwells, as they say of the initiated, in company with the gods.[2]

For Plato, then, as for Coleridge, the essence of philosophy is "the study of death." Idealist philosophy in general is, in fact, centered around an obsession with death. The theme of death is never far below the surface in "Bishop" Berkeley or Kant or their followers, direct or indirect. When, for instance, Kant was trying to shore up the remnants of idealism after Hume's devastations he concentrated on re-establishing the doctrine of the mind as a unique essence and, hence, hypothetically capable of immortality. On the other hand the skeptics, from Confucius to Hume, although close to the idealists on some issues, were skeptical about immortality.

As we consider the views of Coleridge and Plato on death it becomes apparent that, simply put, the emperor has no clothes. Their pious-sounding abstractionist arguments could not withstand five minutes of rational examination. Moreover, their main thrust is egocentric. Although Coleridge is ostensibly contemplating death as a general phenomenon, he is clearly thinking primarily of himself. The vision of the great "soul" escaping annihilation is, we might suspect, inspired more by Coleridge's contemplation of his own death than of Milton's. Socrates' arguments on immortality center on himself – "whither, if God will, my soul is also soon to go" – and on a small elitist group who look to him for personal assurance: "And where shall we find a good charmer of our fears, Socrates, when you are gone?" Such arguments have an air of shrunken self-absorption when we consider the massive afflictions of society and the billionfold cycles of birth and death.

In both Coleridge and Plato, also, the emphasis on the evil of "wild passions" hides, as ever with such obsessions, an underlying anti-feminism. Plato does not attack homosexuality. Indeed, just the opposite.

The classic materialist arguments on death were those proposed by Lucretius: there is no more need to fear death than to fear a sleep "prolonged to eternity;" death is necessary if life is to continue ("The old is always thrust aside to make way for the new, and one thing must be built out of the wreck of another"); we do not trouble ourselves about "the eternity that passed before we were born," so why be obsessed with the eternity still to come? And to these we might add the general materialist emphasis on the predominance of life, expressed, for instance, in the lines of William Ellery Leonard, poet, and editor of Lucretius, on the death of his young bride:

> I built her many a fire for love . . . for mirth. . . .
> (When snows were falling on our oaks outside,
> Dear, many a winter fire upon the hearth) . . .
> (. . . farewell . . . farewell . . . farewell . . .)
> We dare not think too long on those who died,
> While still so many yet must come to birth.[3]

Marx and Engels write even less about death in their works than do earlier materialists because they placed the emphasis not only on life in general but on the social struggles in which they were so totally involved. Like other materialists they doubtless agreed with Lucretius' general views but they saw them within the context of these struggles. We must turn to their letters to find comments on death, usually inspired by personal tragedy.

Although they express the usual and initial human reaction of grief and wonderment, as materialists they see death without illusion and in the context of life. Engels, great though his grief was, believed it was well that Marx had died when he did:

Medical skill might have been able to give him a few more years of vegetative existence, the life of a helpless being, dying—to the triumph of the doctors' art—not suddenly but inch by inch. But our Marx could never have borne that. To have lived with his uncompleted works before him, tantalized by the desire to finish them and yet unable to do so, would have been a thousand times more bitter than the gentle death which overtook him. "Death is not a misfortune for him who died, but for him who survives," he used to say, quoting Epicurus.

The comment of Epicurus was doubtless given special meaning for Marx by the death of his son, Edgar, at the age of eight: "I have already suffered many misfortunes, but only now have I come to know what real grief is." When the German socialist, Ferdinand Lassalle, died in a duel, Engels commented to Marx: "It is hard to believe that such a noisy, stirring, pushing person is now dead as a mouse." When Engels' companion and lover – really wife – of twenty years, Mary Burns, died suddenly, Engels, then 42 years of age, wrote to Marx: "I feel that I have buried with her the last particle of my youth."[4]

Marx and Engels, then, did not try to mitigate the tragedy of death or stifle their natural grief. Nor did they become obsessed with it or with their own inevitable end. This was doubtless because along with their intense involvement with the working class they realized the absurdity of the hypothetical alternative – "the tedious notion of personal immortality," as Engels called it, a concept that was wittily concretized by Bernard Shaw in a remark to the effect of "imagine a million years of Bernard Shaw." When the doctrine of immortality is thus put in personal terms it is evident that it can only be retained by invoking the irrationality of "faith." The emphasis in Marx and Engels is not on the self but on the social group, not on personal dying but on the effects of death on those left behind. Personal death, devastating though the thought of it may be, has to be accepted as necessary for the continuation of life. The survivors mourn, for they are human. The pain of grief is inevitable and can be shattering. But if it is not to become chronic, the survivors must throw themselves into work and living, putting the interest of society above personal feeling, as Marx did after the death of his son and Engels after the death of Mary Burns. The implication of their words and actions is the opposite of that of Coleridge and other theologians, who present death as an all-dominating force that can be faced only by the narcotics of delusion. An obsession with one's own death no doubt seemed a petty thing to Marx and Engels in the light of what Keats called "the giant agony of the world."[5]

What would Marx and Engels have said of the death of Milton (or any other outstanding individual)? They would have said, first, that we simply have to accept the fact that every human life at some point comes to an end and its significance is not to be found in the fact of death but in itself. They would have viewed Milton's death in the perspective of his life-struggle against tyranny, as the leading intellectual in the first antifeudal revolutionary government (that of Oliver Cromwell), as the pioneer author of works advocating divorce (in 1643) and freedom of speech, and as a passionate supporter of revolutionary justice (the execution of Charles I). Even though the narrow religious content of *Paradise Lost* hampered philosophical insight, Milton's central thrust was one of struggle against reaction, with an unconscious sympathy for the rebel Satan. With the magnificent rebel

Samson in *Samson Agonistes,* Milton identified consciously. On the whole Marx and Engels would have seen Milton as a bright and lingering light in the ascending struggle of humanity; his works, both in poetry and prose, lending truth, power and beauty to the struggle.[6]

On the other hand, neither Marx nor Engels attempted to deny the desolating impact of death. The vital Lassalle was, stunningly, dead as "a mouse." Because we have consciousness and live in the intricate webs of human society, it is difficult to fully grasp, even though, as Engels implies, we may "know" it, that the same natural process is at work in the death of a mouse and a human.

Animals have various instincts connected with death. They attempt to avoid death and, indeed, have evolutionary inherited equipment – from speed to protective shells – that helps them to avoid it. Yet they will give up their lives for the good of the group, as when a bird or a monkey on watch cries danger even though it means disclosing itself to the approaching predator. Animal parents will sacrifice themselves for their young. At a certain point, animals seem to accept death. An ape being attacked by others, Jane Goodall noted, will ferociously resist but when it becomes apparent that the fight is hopeless, it simply gives up. When death becomes inevitable animals sense it and, as Lewis Thomas comments, "seem to have an instinct for performing death alone, hidden." Some animals also express grief. Jane Goodall showed – in fact, recorded on film – that a daughter ape could pine away and actually die after the death of her mother. How these and other instincts affect human reactions to death seems to have been little explored. The tendency of animals simply to give up at a certain point and accept death may indicate a genetic basis for suicide. Human grief is apparently not simply a conscious reaction, as it appears to be, but is rooted in instinctive emotional patterns. So, too, perhaps with fears for our own death. James Boswell once asked Samuel Johnson, "is not the fear of death natural to man?" Johnson replied: "So much so, Sir, that the whole of life is but keeping away the thoughts of it." True, the reply indicates an obsession with death similar to that in Coleridge, but, as also with Coleridge, it contains a fragmented truth. Animals seem to "fear" death. And people apparently think more about death, especially their own, than they let on. The sight or even the thought of death can inspire a primitive terror that suggests an instinctive basis. There is, as wakes or even funeral services reveal, a reluctance to face death directly. It is a subject laced with taboos – from hunting-gathering to civilized societies – that mask an underlying, generally shared, fear.[7]

I remember once in a movie theatre in Bloomington, Indiana, seeing a short film on Phineas T. Barnum, the 19th century theatre and circus entrepreneur. The audience was not particularly interested until the film sound track played a speech by Barnum which was, he said, being recorded so that people could hear his voice after he had "joined the great majority." At this point the audience became tense and silent. Clearly everyone there had thought much about death but had probably expressed little of this to others, sweeping the thought aside as painful and returning to the business of life. Barnum's words drew the whole theatre together temporarily in an unspoken communion.

People become uncomfortable even at indirect suggestions of the omnipresence of death; for instance in discussions of the immensity of the universe or of the people of past ages:

> There, like the wind through woods in riot,
> Through him the gale of life blew high;
> The tree of man was never quiet;
> Then 'twas the Roman, now 'tis I.

True, materialism places these matters in perspective. The universe which seems alien to life produced life. The historical past produced the present. But in exploitive societies, in which the dominant-class drive is to isolate death from life and create an obsessive fear of death, this deeper perspective is difficult to achieve.[8]

The fear of death, however, is not, as some seem to believe, a unique phenomenon or a constant but a fear like other fears. Like them it rises and falls with circumstance. When we are depressed it is more evident, when we are upbeat it vanishes, when we face death it can become obsessive. But, as the heroic deaths of revolutionaries – by the millions – show, it is a fear that can be overcome. The fact is that we do not have an emotion or vision adequate to encompass death, Blake's "eternal death," in its relation to life. As with everything else we react on the basis of whatever psychological equipment evolution has given us. We grieve as the animals grieve, although with conscious understanding mingled with instinctive patterns.[9]

Death is in one way the simplest thing in the world to understand and in another the hardest. It is easy to see that life ends for each individual, human or animal, but it is difficult to encompass its significance. One reason for this is that we have no way of directly experiencing death; and we are able to grasp in greater depth things that have a basis in experience. It is easy to "understand" that the end of each life is "forever" but it is difficult to really grasp it or fit it into the perspectives of daily living. It is also difficult to envisage the interaction of history with the biological birth-to-death pattern running through it. We tend to simplify the past, seeing it as a semi-fictional pageant instead of a series of times like our own, vibrant with sun and life, peopled by the once living who are now dead – some of whom have a message for us:

> *Where are your books?—that light bequeathed*
> *To Beings else forlorn and blind!*
> *Up! up! and drink the spirit breathed*
> *From dead men to their kind.*

To Wordsworth with his earthy, country realism the writers of the past are not abstract "spirits" but "dead men," those once living, as we are, now forever gone but with their writings still alive, as now with Wordsworth himself. When we think of posterity we tend also to think, as the word itself indicates, in an abstractionist way, not of actual people being born, "While still so many yet must come to birth," and dying.[10]

We can become more aware of the reality of the past through simple daily detail. Percy Bysshe Shelley, eloping with Mary Wollstone-craft Godwin on July 28, 1814, hired a "small boat to convey us to Calais." "The evening," he recorded at the time, "was the most beautiful; the sands slowly receded: we felt safe; there was little wind, the sails flapped in the flagging breeze." True, we need to perceive the great movements of the past, in 1814 of The Napoleonic Era, the age of Metternich and Nelson, of Beethoven and Goya. But it was not just an "era." It was the scene of everyday human activity – of gently flapping sails above two young lovers off the beach at Dover on the evening of July 28, 1814.[11]

When I was writing background essays and notes for the four volumes of *Shelley and his Circle* that I edited, I was always turning up the names of people of whom I had never heard. This sent me to the London city directories of the time, and as I surveyed the seemingly endless alphabetized rows of names now utterly forgotten it became apparent how distorted is our view of the people of the past. Particularly in class-exploitive societies, we are directed only to a kind of hierarchy of the gods, the few – writers, generals, statesmen, actors and so on – whose names have survived. The rest fade into an amorphous backdrop. Reading such directories also brings the theory of immortality down to earth. Are we to suppose that all these people exist somewhere as eternal wisps of mind or as bodies that lie awaiting "resurrection" and "judgment" for perhaps millions of years? Are some family members then to go eternally to Hell and others to Heaven?

As we grow older, it becomes somewhat easier to envisage the birth-death cycle of the ages because we have experienced something of it. We have lived through the reality of past periods (that are often now presented in unreal and distorted forms). We can see old people as they must have been as children and we can see children as they will be when old. We begin to grasp at the essence of the immense life-death stream of life.

As we survey the panorama of evolution, it becomes apparent that life is not primarily an individual but a group phenomenon. When Schaller went back to Africa after a brief three years, he found considerable changes in his gorilla groups. Some gorillas had died, others grown old, groups had broken and mingled; of one group of twenty-one only ten were left. Yet the groups as groups continued, the whole continued. With animals shorter-lived than the apes, such as butterflies, the omnipresence of the group and the ephemeral nature of the individual is even clearer. If it is less clear in human society, this is because we live longer than most animals and are conscious of our existence. But it still holds true. And for a Marxist it supplies a basis for a view of life and death that was hidden from Lucretius and other earlier materialists.[12]

Lucretius had, of course, no concept of objective social evolution and its relation to progressive collective activity. He did not, indeed, could not, perceive that those fighting collectively for social advance are vitally supported by the knowledge that their group (organization) will continue after their death and they can put the life-death cycle into this perspective. Especially is this true of the members of a Marxist-Leninist party, for such a party projects from the present a society of a radically new kind, based not on the economic exploitation of the mass of humanity but upon a mass-democracy economically geared to its needs. The first stage of such a society existed for some decades in the Soviet Union and elsewhere, and doubtless will again, as a harbinger of a world communist commonwealth.

Our reaction to death is not, as Lucretius implies, primarily intellectual. As Marx early emphasized, we are a mixture of thought, emotion and "practice," and each affects the other. All play a part in philosophical matters including our reactions to death. The easing of grief is not brought about mainly by reasoning but by group actions, religious or secular, family or otherwise. Radical social groups can help those who have suffered a loss through death by encouraging activity and combining it with philosophical discussion of the views on death of Lucretius and other materialists. This will not, of course, eliminate grief or the fear of death, but it can mitigate them and help to rebuild the lives of the survivors. Those who flush the narcotics of delusion out of their minds may lose a comforting crutch but they are also no longer racked by the terrors of Hell and Sin or the useless torment of sustaining Faith in the face of an obviously intractable reality.

The philosophical implications of animal society have, as I have noted, been largely missed. Even those who "accept" evolution may not see it as part of their living or integrate it with their everyday thinking. Here our household pets can help us bring things down to earth. A few minutes ago our cat leapt up beside my typewriter. He did this because he needed contact with another living creature – me (in lieu of other cats) – and he complained when I lifted him down. He sought this contact, however, not because "animals are like us" but because we are like the animals. Although our evolutionary ancestors long ago branched up and away from those of the cat, we are still in some ways the same. The cat, like us, can feel fear and joy, love and hostility. Humans and cats can respond together to these feelings – for instance in play – even though they are essentially unconscious in the cat and blend with consciousness in us. It is, in fact, mainly because of this emotional reciprocation that we can feel close to pet cats and dogs and other animals. They are, so to speak, still within us even though we transcend them.

When we realize this, it is an exhilarating thing for we can see that we and our pet animal alike embody the world life process. Yet it is also depressing, for it brings us up against our own mortality. We die as the cat dies. Death is strewn like kelp over the waves of evolution. We begin also to perceive the artificial situation of the cat, which should be in a jungle with its young instead of being a lone creature looking for affection in an alien setting. Yet, is it really "looking" for anything? Does it really "complain"? The more we observe the cat, the more we realize that in spite of its seeming consciousness it is a primarily bio-psychological mechanism. For instance, if we watch as it is about to leap on a thin balcony railing (or the edge of a typewriter stand), it seems to be judging its distance and the nature of the landing area. In fact, it is acting automatically, the source of its actions lying primarily in the cerebellum (which biomechanically regulates such things as balance). Yet in seeing this, we begin to see more of ourselves, to see that our conscious thoughts and feelings arose as compounds of a blind natural process, one of whose stages we can perceive in the cat. In short, although it is difficult to visualize the mental world of the cat because it is both complex and non-conscious, we can see how such complexity when carried a stage or two further in the apes trembles on the brink of consciousness.

The cat shows us also that our consciousness arises from a compound of non-conscious elements. We can see this also in dreams, for in dreams a mechanism in the brain which is not "us" takes over and arranges thoughts, emotions and images into patterns over which "we" have no control. This is, when we first consider it, a disturbing thing for we see that what we had considered as an inviolable whole, namely our being, is actually a composite of parts built up by evolution.

The cat represents life on the planet before we evolved, a planet of creatures like cats, for many millions of years, and before that of creatures like lizards. These things too, we can learn to see in the world around us. One day I was on the beach – off West Basin Road at Gay Head on Martha's Vineyard – observing, as I had for many years, the swift tern swooping for fish, when it occurred to me that this spectacle must have been going on for tens of millions of years, long before there were any humans around to watch it.

There seems to be little in the works or letters of Lenin on death. His life and writings were so consumed by the working class struggle, first for overthrowing capitalism and then for laying socialist foundations, that he considered death, when he considered it at all, in the light of this struggle. During the turmoil of the 1905 revolution he wrote: "Contempt for death must spread among the masses and thus secure victory." When in *"Left-wing" Communism, An Infantile Disorder* (1920), he discussed the components of a revolutionary situation he included among them the fact that "the majority of the class-conscious" workers must "fully understand that revolution is necessary and be ready to sacrifice their lives for it." The implication in both comments is that people have a natural fear of death and this must be conquered if a revolution is to succeed. In another passage in *"Left-wing" Communism,* he wrote that although the modern bourgeoisie might in a reactionary "frenzy" slaughter "thousands, and hundreds of thousands" of revolutionaries and set back a particular revolution, in time "life will assert itself" and the revolution triumph.[13]

Lenin, then, recognized the intrusion of biological factors into the historical stream. The slaughter of communists may not stop a revolution but it can delay it. Similarly with war. A second war cannot follow immediately after a really devastating war for it is necessary to wait for a new generation – especially its young men – to be born and bred. Individual death can also have historical effect. The death of a political leader such as Stalin or Mao cannot change the basic direction of events but can certainly modify it. The life cycle, from youth to age, as it weaves its way through the patterns of history, has an effect on those patterns. Revolutions generally represent a blending of younger and older revolutionaries, the younger supplying most of the energy and physical valor, the older most of the basic guidance. The young Party members, Stalin noted in 1939, "develop and acquire knowledge so rapidly, they press forward so eagerly, that the time is not far off when they will overtake the old fellows." Moreover the old are "already partly going out of commission owing to the operation of the laws of nature." We might note, too, that without the instinct for self-sacrifice for the good of the group, the willingness to sacrifice one's life for a revolutionary cause that Lenin speaks of would not exist.[14]

Because Marxism regards history as evolutionary, it sees a philosophical significance in it that other world-views do not. For instance, the slave revolts of slave-commercial Italy, and the peasant revolts in feudal Britain or China, are not meaningless incidents in a cyclical pattern but parts of a historical spiral. The challenge to oppression by Spartacus and Wat Tyler, Ali of Basra and Hung Hsiu-chuan, by Cromwell and Robespierre, by Toussaint L'Ouverture and Mary Wollstonecraft, by Paine and Byron, has a larger social meaning than they were aware of, for it was part of the historical ascent of humanity to a world communist commonwealth. This ascent is an immense process but in general each human life has been part of it, building segments of the socio-economic patterns that have laid the base for the advance of society from hunting-gathering to feudalism, to capitalism, to socialism. Many millions have also taken part in the socio-political struggles: from, for instance, the early Chinese peasant revolts to the Vietnamese revolutionary war against French and U.S. imperialism, that integrate with the structural economic patterns to drive society forward. In short, without the struggle and toil of humanity for thousands of years, there would have been no industrial capitalism, no socialism and no possibility for a future communist world. Thus, there has been a collective purpose to life derived from sharing a universal effort, a purpose arising not from supernatural benevolence but from the nature of society. In the convulsive decades just ahead, of biosphere destruction, this sense of collective purpose will become a universal bond as humanity struggles for survival and turns unprecedented disasters into spurs for further socialist advance.[15]

As capitalist expansion in the late 18th century produced the American revolutionary war and the French Revolution, some began to perceive that history was pointed upward and became conscious of a purpose in their lives derived from this fact:

> *Scatter, as from an unextinguished hearth*
> *Ashes and sparks, my words among mankind!*
> *Be through my lips to unawakened earth*
>
> *The Trumpet of a prophecy! O, wind,*
> *If Winter comes can Spring be far behind?*

Shelley's "Winter" symbolized the reactionary bastion of post-Napoleonic Europe, Spring the egalitarian Utopia to follow. Shelley saw his "words" as helping to shape that future.[16]

As the industrial revolution continued, the historically new class of (paid and literate) industrial workers produced a new world outlook:

> An entire change in society—a change amounting to a complete subversion of the existing "order of the world"—is contemplated by the working classes. They aspire to be at the top instead of at the bottom of society—or rather that there should be no bottom or top at all.

So wrote the British Chartist leader Bronterre O'Brien in 1833, when Marx was but fifteen.

But it required Marx with his combination of utter absorption into the working class and his sweep of historical vision to shape the new ideology into one of scientific analysis and projection:

> Between capitalist and communist society lies a period of revolutionary transformation from one to the other. There corresponds also to this a political transition period during which the state can be nothing else than the revolutionary dictatorship of the proletariat.[17]

In the present century we see a new phenomenon, namely a new mass sense of purpose, which arose when, for the first time in history, people created revolutions that gave them economic control of society, first in the USSR and then in China and other socialist countries.

At present this sense of purpose and the society behind it face two threats. The first comes from the severe setbacks socialism has received in the Soviet Union and the countries of east Europe, from Poland to Romania. At first these setbacks were gleefully presented in the capitalist media as "the collapse of communism." But it soon became apparent that political change did not automatically bring with it socio-economic change. Basic socialist economic and other structures continued to exist, particularly in Russia and some other former Soviet republics. It has also already become apparent that capitalism, represented as an endless vista of *Rich and Famous* television glamor, has instead brought economic crises, social impoverishment and bourgeois nationalist slaughter. Instead of instant communist collapse, we are now witnessing a growing class struggle, particularly evident in Russia, between the forces of capitalism and socialism. When we note that at the same time the corporate-capitalist world is itself caught up in a worldwide socio-economic crisis, it appears that beneath the seemingly chaotic "sound and fury" on the surface, the deeper historical currents are still moving, as they were in the first half of the century, towards socialism. Massive and rapid change is in the offing. It is, as Palme Dutt pointed out in a review of Sidney and Beatrice Webb's *Soviet Communism: A New Civilization?* (1935) easy to see events when they are spelled out on the blackboard of history; the thing is to see them, as Lenin did, before they happen.

The second threat I discuss in Appendix II, "Socialism and Planetary Destruction," in my book *Marxism: A Living Science* (New York, 1994). The subheads give the drift of the presentation: Global Warming; Ozone Destruction; The Way Out.

The corporate media have for some years spread confusion about global warming, including – a specialty of the *New York Times* – raising doubts on the validity of the theory: "there are enormous uncertainties." But although scientists acknowledge some uncertainties, these are matters of detail. Of the fact of global warming, the evidence has been clear for the past ten years or more. And recent evidence indicates that it is already in operation. Moreover it is not, as the corporate media contends, a phenomenon of limited scope. Unless it is halted it will destroy humanity, turning the planet into a furnace like Venus.

Behind the media's obfuscations lies the simple fact that corporate capitalism's profits basically arise from the burning of the fossil fuels that produce carbon dioxide, the gas that regulates Earth's temperature.

Life on land became possible only when some 400 million years ago an ozone layer was formed in the stratosphere and began to filter the sun's lethal streams of ultraviolet radiation. In recent decades the CFCs – essentially chlorine – and other chemicals have begun to destroy this layer, which will result not only in an epidemic of skin cancer but in the destruction of the plankton at the root of the oceanic food chain, and of such basic crops as rice and soybeans.

To the destruction of the upper atmosphere we have to add the pollution of the lower atmosphere, ocean and land – particularly nuclear pollution – all in a gigantic dialectical swirl, with upper atmosphere destruction as its central component.

It is now becoming ever more apparent that capitalism, with its profit-driven anarchy of production, cannot build a clean energy economic base or, indeed, take any measures that would threaten its interests. For this socialist mass-oriented planned economy is needed.

In the next few decades we can expect to see a confluence of two struggles, that for the salvation of humanity and that for socialism. Both will certainly be conducted amid unprecedented natural disaster and social turmoil.

As we have seen, our instincts for survival underlie morality. We exult in actions that we perceive to be in the group interest and condemn those that oppose it. These forces blend with the socialist-society sense of purpose and give it special vitality. They also underlie the growing sense of purpose apparent in the masses in third world countries as their anti-imperialist struggles increasingly reveal their actions to them as moral and those of their exploiters, external and internal, as immoral (anti-group). These insights, consciously or unconsciously, become part of the political struggle. So, too, with Marxists in monopoly-capitalist nations, continuing a tradition begun in the 19th century with its great labor struggles, economic and political.

Those who see life as supernaturally created, directly or remotely, necessarily place a primary emphasis on death, for birth and growth are taken for granted as a kind of "gift" from God. But when we see that everything arises from natural forces and from nothing else, life is seen as primary. Life came into being from one complex of material interactions and evolved from another. It could just as easily never have come into existence or never produced mammals or primates. The gossamer vibrances of the nucleic acids that shape the fabric of life are rooted in matter and matter alone. The essence of our being lies in the vitality provided by the oppositeness of matter visible in the clash of atoms. As infants become children, innate instinctive patterns develop in response to social activity and blend with consciousness. If social activities ceased; if for instance, parents, particularly mothers, ceased to devote part of their lives to their offspring, the human race would perish. The patterns of human life, unlike those of reptiles, could not develop on their own.

Although it seems likely that lower forms of life exist on other planets in the universe, when we consider our evolutionary course and its roots both in the cosmos and on Earth, it seems infinitely improbable that human life could exist elsewhere. We are almost certainly alone in the universe. There was no miracle involved in the emergence of life on the planet and no miracle will avert its destruction. When we consider the trillionfold chance combinations of sperm and ova over the centuries that were needed for the making of each individual it becomes clear that it is also virtually infinitely improbable that any one of us should exist. The extraordinary thing is not that we die but that we are here, collectively and individually.

Religious thinkers and philosophical idealists like to assert the "oneness" of humanity with Nature, but they see this oneness as arising from supernatural forces. Plato fantasized it as a spiritual union between God (the One), who suffuses nature, and the human mind (or soul). And religious-minded people in general feel the link between themselves and nature to be "divine." In fact, such fantasies obscure the actual links between nature and humanity, some of which have long been noted by materialist philosophers. Lucretius implied that such links were based on the common atomic nature of matter, body and mind ("refined" atoms); but he saw these relationships only in a general way and in quasi-evolutionary terms. So, too, with Diderot. Then the 19th century discovery of the cell and the theory of natural selection began to reveal the actual links between the varied forms of life from bacterial to human. Marx and Engels, like Darwin, saw the connection between "nature" and people as arising from evolution. But still little was known of the nature of living matter or of the brain. Contemporary science, however, has begun to uncover the specifics, from elementary particles to molecules, from animals to humans, so that today we can have a deeper understanding of nature and people than was possible in the past. On the other hand, non-Marxist scientists, even as they uncover specific connections, blunt their significance by placing them solely within the boundaries of their specialties, unaware of their relation to the general (dialectical) processes of nature. And even these specifics they sometimes distort by mysticism: "I see in the brain all the beauty of the universe and its order – constant signs of God's presence."[18]

By combining such diverse but nevertheless related phenomena as star explosions and the evolution of the brain it becomes apparent that the intense identification with living nature we find in such poems as Blake's *The Tiger,* have evolutionary substance of which their authors were unaware but had some sense of:

> *Tiger! Tiger! burning bright*
> *In the forests of the night . . .*
> *In what distant deeps or skies*
> *Burnt the fire of thine eyes?*

The elementary particles of life and light were born in the big bang.

The carbon, oxygen, iron and other minerals in our bodies came from the explosion of stars many millions of years ago. Our brains run on the interactions of electrons and atomic clusters. The fragmented elements of life exist on meteors and in cosmic dust. Shelley was able to identify with the skylark partly because – although he did not know it – both were the products of evolution. So, too, Swinburne and his "sister," the swallow.[19]

We are part of living matter and its unwinding spirals. Our growth process – from conception on – is not only basically the same as that of the animals, it illuminates the stages of our evolution as well as theirs. We can now see in colored motion pictures the earlier evolutionary fetal forms from which we arose. The past within us is being blended by science ever more intimately with our living and thinking.

In Engels' day, the electric aspect of the atom was unknown, and it was just being uncovered when Lenin wrote *Materialism and Empirio-Criticism.* Nor was it known that the electric "reaction" has roots in the general oppositional nature of elementary particles. We can now see that our life essence is not just one with living matter – as vividly revealed in films of the agitated dartings of spermatazoa or the growth of the fertilized ovum – but also with the clash of atoms and particles; with clouds, wind and lightning as well as with plant or animal life: "The force that drives the water through the rocks drives my red blood." We can see that the restless, conflictive nature of life – bursting through the Arctic tundra and cracks in the sidewalk – arises ultimately from the electric clash of atoms.[20]

We have a sense of identity not only with the growing forces of nature but with their spirals of decline and death:

> *Cherished by the faithful sun,*
> *On and on eternally*
> *Shall your altered fluid run,*
> *Bud and bloom and go to seed,*
> *But your singing days are done;*
> *But the music of your talk*
> *Never shall the chemistry*
> *Of the secret earth restore.*

So Edna St. Vincent Millay on the death of a young woman friend. Or Shelley on the death of Keats:

> *the intense atom glows*
> *A moment, then is quenched in a most cold repose.*[21]

Modern advances in science, it is frequently argued, remove the picture of reality revealed by our unaided senses increasingly away from that revealed through the tools of science. In some ways this is true. Our bodies are, on one level, mainly space laced by particles and atoms in electric impulses, on another level patterned bundles of macromolecules and cells. Light consists of a massive barrage of individual particles. Behind the mind lies the neuron.

But the opposite is also true. No matter how many varieties of particles are discovered they all, except for the photon, the most elementary of all, have opposites, and the photon is said to encompass its own opposite. No matter how many processes of nature are unveiled – from pre-atomic matter to evolution – they all change under the spur of these opposite entities and develop by converting the old into the new simply through quantitative-arrangement changes. So, too, with history. So, too, with our thinking and feeling. The more specific the discoveries, the deeper our understanding of the general nature of the universe and ourselves becomes.

In this regard we might recall Lenin's simple but profound comment that all the levels of nature – the foam on the river's surface as well as the deep waters beneath – are alike aspects of reality. Therefore, we can grasp something of the general character of aspects of nature beyond our senses' unaided penetration. The extraordinary degree to which this is possible is illustrated by Lucretius who, with almost none of the tools of science at his disposal and relying primarily on observation and reasoning, came up with views generally similar to those revealed by science today, for example: "But multitudinous atoms, swept along in multitudinous courses through infinite time by mutual clashes and their own weight, have come together in every possible way and realized everything that could be formed by their combinations."[22]

The argument, advanced even by some scientists, that we can never know the truth about nature is at bottom, even if advanced as skepticism, religious mysticism. The fact is that we are more and more penetrating surface appearance and discovering underlying process, not only in nature but in society and ourselves. There is no hidden "something else," no supernatural beyond the natural. This was argued by the early Greek materialists and perhaps the Indian materialists before them and has become vividly apparent as science has filled in the picture. So, too, as Marx discovered, with society. Society, which appears to respond solely to direct human endeavor, responds on a deeper level to social structures that have their own dynamic. The only basic forces shaping society and people are those arising from social and natural interactions. Neither the Hand of God nor Fate guides human "destiny." The human psyche is not akin to "the Divine" but is rooted in the animals and nurtured by society.

It is not, of course, in the interest of an exploitive ruling class, with its inherent need to spread debilitating superstition, to encourage the development of a dialectical materialist viewpoint. Although natural science may be needed for advancing the economic interests of the class, its philosophical implications are routinely distorted. Social science – Marxism – is not only not needed by such a class but is abhorred, and to distortion is added the persecution of its exponents to whatever degree the balance of class forces allows, from wrecking careers to imprisonment, torture, execution and mass slaughter.

Our brains were evolved by natural selection not to indulge in "pure reason" but to hunt, to fight, to mate, to work in groups, follow leaders, nurture and protect offspring. Integrated with this development were the perceptions and thinking processes needed for such tasks. We combine living with reason, but living is the basic motivating force. And this is just as well. True, nature and society are not quite what they seem and we are part of an interaction of atomic, molecular and social processes, but our understanding of this is prevented from becoming dominant in our living by the fact of living itself. As I put it in an earlier work, we live on our "life juices." Otherwise, life would be impossible. We accept, in thought and activity, the world as it appears even though we acquire knowledge of its basic aspects and this vitally affects our thinking and living.

We accept the growth of a baby not primarily as a neurological but as a human phenomenon, and we respond primarily not with scientific analysis but with love or joy or fear. So, too, with our living in general, in our relations with people, society, animals and inanimate nature. In short, we necessarily live primarily on our own level of reality. But it is scientific analysis that gives understanding and power. Knowledge increases our understanding of the connections between our own and other levels of reality and enables us to utilize both natural and social process, and so live more deeply. Here the Marxist world-view is the supreme blender, driving us to integrate our living and thinking in the ascending struggles of the present: "The point is to *change* it."[23]

Notes

Chapter One

1. For general accounts of human biological and early social evolution, see, for instance, Richard E. Leakey and Roger Lewin, *What Makes Us Human?* (New York, 1993); Donald Johanson and Maitland Edey, *Lucy: The Beginnings of Humankind,* New York, 1981; and *Ancestors, Four Million Years of Humanity,* American Museum of Natural History (New York, 1984). See also the *New York Times,* Feb. 3, 1992, on *homo erectus* remains of possibly 1.6 million years ago in Asian Georgia; Feb. 14, 1989 on fully human remains of 90,000 years ago; and Nov. 18, 1993, on more "Lucy" finds.

2. Robert H. Lowie, *An Introduction to Cultural Anthropology* (New York, 1955), p. 327.

3. Ronald M. Berndt and Catherine H. Berndt, *The First Australians* (New York, 1954), pp. 129-130.

4. Kenneth Neill Cameron, *Humanity and Society: A World History* (New York, 1977), pp. 35-46; *New York Times,* Dec. 16, 1986.

5. On farming society as a stage in historical evolution, see my *Humanity and Society,* Chapter III. In that account I did not give enough emphasis to the tribal aspect of such societies, which still continue in some areas of Africa and Asia. For a vivid account of the emerging class struggles and bitter inter-tribal conflicts in such societies in their later stages, see the film *Chaka Zulu,* shown from time to time in the United States on Public Television.

6. Stanley Kramer, *History Begins at Sumer* (New York, 1959), p. 125; *Gilgamesh,* trans. William Ellery Leonard, quoted in Cameron, *Humanity and Society,* pp. 77, 78; H. and H. A. Frankfort, John A. Wilson, Thorkild Jacobsen, *Before Philosophy: The Intellectual Adventure of Ancient Man* (Pelican Books, 1949), pp. 226, 184. See also: *The Epic of Gilgamesh,* trans., ed. by N.K. Sanders (Penguin Books, 1980) with its scholarly Introduction.

7. *Before Philosophy,* p. 66. Although particular views and attitudes are often a direct reflection of class interests and obvious as such to the class involved, the formation of a total class ideology is a complex historical process. Some of the ways, often unconscious, in which it is formed I discuss in my *Marxism, A Living Science* (New York, 1994), pp. 137-149.

8. *Before Philosophy,* p. 155; Kramer, p. 122. As Kramer notes, the comment indicates "a cer-

211

tain degree of class consciousness."

9. Frederick Engels, *The Origin of the Family, Private Property and the State* (New York, 1942), p. 58; Sheila Rowbotham, *Hidden from History: Rediscovering Women in History from the 17th Century to the Present,* (New York, 1976). See also: Cameron, *Marxism, A Living Science,* Ch. 6.

10. *The Wisdom of China and India,* ed., Lin Yuyang (New York, 1942), p. 15.

11. Ibid., pp. 62, 63. See also Ralph Waldo Emerson's rendition of some of these ideas in his poem *Brahma:*

> *If the red slayer thinks he slays,*
> *Or if the slain thinks he is*
> *slain,*
> *They know not well the subtle*
> *ways*
> *I keep, and pass, and turn*
> *again.*

12. Will Durant, *The Story of Civilization,* "Our Oriental Heritage" (New York, 1954), p. 420.

13. *World Bible,* ed., Robert O. Ballou (New York, 1956), p. 119.

14. Ibid., p. 547; *Wisdom of China and India,* p. 584; *World Bible,* pp. 556, 503, 505.

15. For a further brief examination of Greek society, see my *Humanity and Society,* pp. 160-170.

16. *Phaedo, The Works of Plato,* ed., Irwin Edman (New York, 1928), p. 139.

17. "Physics," Aristotle, *Selections,* ed. W. D. Ross (New York, 1927), pp. 100-101: "Politics," ibid., p. 293; "Categories," ibid., p. 3;

"Metaphysics," ibid., p. 56; "Categories," ibid., p. 7. "Metaphysics," *The Portable Greek Reader,* ed. W.H. Auden (New York, 1955), p. 198; "Categories," *Selections,* p. 7; "Metaphysics," *Greek Reader,* p. 198.

18. Benjamin Farrington, *Greek Science, Its Meaning for Us* (Penguin Books, 1955), p. 162.

19. *Greek Reader,* pp. 71-74. Marx's thesis was titled *The Distinction Between the Democritian and Epicurean Philosophies of Nature.*

20. Marx read Lucretius as a university student and praised his "bold thundering song" with its message of "a nature without god and a god aloof from the world." *Marx, Engels, On Literature and Art* (Moscow, 1978), pp. 208, 209, quoting Marx's Notebooks on *Epicurian Philosophy.* The philosophy of Epicurus (341-270 B.C.), one of the founders of the materialist school whose views Lucretius developed, formed a major part of Marx's doctoral dissertation. See *Karl Marx, A Biography* (Moscow, 1973), pp. 29-31. Alban D. Winspear, *Lucretius and Scientific Thought,* (Montreal, 1963), Winspear, a fine Marxist and scholar, was a close friend of mine at the University of Wisconsin (Madison) in the late 1930s.

21. Lucretius, *The Nature of the Universe,* trans. R.E. Latham (Penguin Classics, 1951), pp. 135, 37, 250, 31, 182, 183-184. The title in Latin is *De Rerum Natura (On the Nature of Things).* Winspear, op. cit., pp. 1-15.

22. *Nature of the Universe,* pp. 31, 182, 43, 90.

23. Ibid., pp. 197, 85, 245, 116, 101, 125.

24. Ibid., 100, 188; Cameron, *Humanity and Society,* pp. 201-203.

25. Sophocles, *Antigone,* lines 381, 389, 363-365.

26. The Nicene creed, based on still earlier creeds, was drawn up by the Council of Nicaea (near Constantinople), and is the only creed generally adopted by Christian churches. It is recited at Holy Communion in the Church of England.

27. Will Durant, *The Story of Civilization,* "The Age of Belief," (New York, 1960), p. 152; Alfred E. Housman, *Last Poems;* Thomas Aquinas, *On the Truth of the Catholic Faith* (New York, 1956), p. 275.

28. Roger Bacon, quoted in Durant, *op. cit.,* p. 138.

29. Francis Bacon, *Novum Organum,* XCVI, XIX.

30. Benjamin Farrington, *Francis Bacon, Philosopher of Industrial Science* (London, 1951), pp. 69, 101, 151-152.

31. Ibid., pp. 78, 79, 136; Marx, Engels, *The Holy Family,* 1844, (Moscow, 1975), pp. 150-151 (passage on Bacon by Marx).

32. "Discourse on Method," Descartes, *Selections,* ed. Ralph M. Eaton (New York, 1927), pp. 29, 30.

33. *Theological-Political Treatise, The Chief Works of Benedict de Spinoza,* tr., R.H.M. Elwes (London, 1889), I: 19, 5.

34. John Locke, "Essay Concerning Human Understanding," Locke, *Selections,* ed., Sterling P. Lamprecht (New York, 1928), p. 206; ibid., pp. 175, 188; 50, "Toleration"; 73, "Liberalism in Politics."

35. George Berkeley, "Principles of Human Knowledge," Section 29, Berkeley, *Essay, Principles, Dialogue,* ed., Mary Whitton Calkins (New York, 1929), pp. 139, 143-144.

36. David Hume, "A Treatise of Human Nature," III, 4, Hume, *Selections* (New York, 1927), pp. 76, 77. Hume, living in an age of English invasion of Scotland, perhaps took a certain glee in undermining the British philosophers.

37. Hume, "Dialogues Concerning Natural Religion," ibid., p. 324; "An Inquiry Concerning Human Understanding," ibid., p. 156.

38. Paul H.T. Holbach, *Le Christianisme devoile,* quoted in Virgil W. Topazio, *Holbach's Moral Philosophy,* Geneva, 1956; Holbach, *The System of Nature,* trans., H.D. Robinson (Boston, 1839), I, 15; ibid., pp. 33, 39.

39. Engels, *Feuerbach, Selected Works,* III, 349-350; *Diderot, Interpreter of Nature, Selected Writings,* ed. Jonathan Kemp, trans., Jean Stewart and Jonathan Kemp (New York, 1943), pp. 72, 136, 28, 74; *System of Nature,* I, 176; *Diderot,* pp. 57-58; *System of Nature,* I, 119.

40. Erasmus Darwin, *The Temple of Nature* (London, 1803), "Additional Notes," p. 45; Erasmus Darwin, *Zoonomia* (London, 1796), I, 509, 139; Percy Bysshe Shelley, *Complete Works,* ed. Roger Ingpen and Walter E. Peck (New York, 1929), VI, 50.

41. Shelley, "Prometheus Unbound," IV, 246-248; Shelley, "A Refutation of Deism," *Complete Works,* VI, 49, 50.
42. Kant's *Prolegomena,* quoted in: Theodore Oizerman, *Dialectical Materialism and the History of Philosophy* (Moscow, 1979), pp. 146-147.
43. Raymond Plant, *Hegel* (Bloomington, Indiana, 1973) pp. 136, 137, 139, 142. Hegel apparently took the terms thesis, antithesis and synthesis from Johann Fichte (1762-1814). Richard T. De George, *Patterns of Soviet Thought* (Ann Arbor, Michigan, 1966) p. 15.
44. Plant, p. 141; De George, p. 15.
45. Hegel's *Logic,* quoted in David Guest, *A Textbook of Dialectical Materialism* (New York, 1939), p. 48.
46. Quoted in: Loren R. Graham, *Science and Philosophy in the Soviet Union* (New York, 1972), p. 53; G.V. Plekhanov, *Fundamental Problems of Marxism* (Moscow, 1974), p. 35; quoted in Frederick Engels, *Dialectics of Nature* (London, 1940), p. 30.
47. Bronterre O'Brien, the Chartist leader, quoted in Cameron, *Marxism, A Living Science,* p. 14. Joseph Dietzgen (1828-88), a tanner, published a materialist oriented philosophical study *Das Wesen der menschlichen Kopfarbeit* (*The Nature of Man's Mental Activity*) in 1868, a work read by Marx, Engels and Lenin. On Richard Carlile, n. 1, next column, para. 2.

Chapter Two

1. *The German Ideology,* Karl Marx and Frederick Engels, *Selected Works* (Moscow, 1973), I, 25, 47. These and other ideas in *The German Ideology* were later developed in the *Communist Manifesto* and Marx's general statement in his Preface to his *Critique of Political Economy* in 1859. (Ibid., pp. 502-506.) Marx and Engels attempted to get *The German Ideology* published but they did not succeed. It was first published in the U.S.S.R. in 1932. Marx commented in his 1859 statement: "We abandoned the manuscript to the gnawing criticism of the mice." Ibid., p. 505; *Karl Marx: A Biography* (Moscow, 1973), pp. 104-105. Karl Marx, *A Critique of the Gotha Program* (1875), (New York, 1933), p. 29. Kenneth Neill Cameron, *Marxism, A Living Science,* pp. 10-11, 15, 18-19, 107.

There are, however, materialist elements in early working-class anti-clerical writings. The most widely known early British working class deist and anti-clerical was Richard Carlile (1790-1843), a tinner by trade, who was imprisoned and persecuted for his writings in his *The Republican* and other publications and for publishing Paine's works. (See G.A. Aldred, *Carlile, Agitator, His Life and Times,* [Glasgow, 1941] and G.D.H. Cole, *Carlile,* [London, 1942].)

On the three-cornered class struggle (in Europe) see Frederick Engels, *Herr Eugen Düh-*

ring's Revolution in Science (Anti-Dühring), (New York, 1939), p. 24. O'Brien, quoted in Asa Briggs, *Collected Essays* (Champaign, Illinois, 1985), I, 20. The Communist League was first called The League of the Just. One chapter of *The German Ideology* was published in 1847.

2. Thomas Paine, *The Age of Reason, The Life and Works of Thomas Paine,* ed., William M. Van der Weyde (New Rochelle, New York, 1925), VIII, 4.

3. Marx to Engels, Jan. 8, 1868, Karl Marx and Frederich Engels, *Correspondence, 1846-1895* (London, 1934), p. 234.

4. Engels, *Anti-Dühring,* Preface, p. 13 (1885).

5. Frederick Engels, *Socialism: Utopian and Scientific, Selected Works* (Moscow, 1973), III, 131. This passage appeared first in *Anti-Dühring* (p. 31, in a different translation):

> In both cases modern materialism is essentially dialectical, and no longer needs any philosophy standing above the other sciences. As soon as each separate science is required to get clarity as to its position in the great totality of things and of our knowledge of things, a special science dealing with this totality is superfluous. What still independently survives of all former philosophy is the science of thought and its laws – formal logic and dialectics. Everything else is merged in the positive science of Nature and history.

6. Frederick Engels, *Feuerbach* (New York, 1941), Appendices, p. 66.

7. Ibid.; *see also* Engels to Marx, July 14, 1858, *Correspondence,* p. 113.

8. Marx to Ferdinand Lassalle, Jan. 16, 1861, *Correspondence,* p. 125; *Feuerbach,* Appendices, p. 67. Marx sent a copy of *Capital* to Darwin, who replied (on Oct. 3, 1873):

> I thank you for the honour you have done me by sending me your great work on *Capital:* and I heartily wish I was more worthy to receive it, by understanding more of the deep and important subject of political economy.
>
> Though our studies have been so different, I believe that we both earnestly desire the extension of knowledge and that this in the long run is sure to add to the happiness of mankind.
>
> (*Karl Marx, A Biography,* [Moscow, 1973], p. 320)

9. Engels, *The Part Played by Labor in the Transition from Ape to Man, Selected Works,* III,* 73-74: "how unerringly the fox makes use of its excellent knowledge of the locality in order to elude its pursuers." This, Engels considered a "planned action." It is not, however, consciously planned but arises from a combination of inherited behavioral patterns with learning and memory. (See Chapters Five

---

Selected Works hereafter as *Selected,* volume I, II or III.

and Seven above.)

10. *Selected,* III, 66-67, 74; Remarks on Haeckel by Stephen Jay Gould in a "Workshop Discussion" on "Karl Marx on Science and Nature," in *Science and Nature* (1978, No. 1), p. 12. Ernst Heinrich Haeckel (1834-1919) was the leading German Darwinian, best known for his materialistic *The Riddle of the Universe* (1899). Engels refers to his *History of Creation* (1868) several times in *Dialectics of Nature* and was indebted to him for the general theory that individual development (from conception on) reflected evolutionary development. John Gribbin and Jeremy Cherfas, *The Monkey Puzzle* (New York, 1982), Chapter One, "One Percent Human"; Donald Johanson and Maitland Edey, *Lucy: The Beginnings of Humankind,* (New York, 1981).

In "The Part Played by Labor," (*Selected* III, 67) Engels argues that "the human hand" has been "perfected by hundreds of thousands of years of labor." This implies the inheritance of acquired characters – a theory accepted in Engels' day by Darwin and others – and hence is incorrect. Some modern Marxists continue to blindly follow Engels on this and other long-discarded views.

11. Karl Marx, *The Economic and Philosophical Manuscripts of 1844,* ed., Dirk J. Struik (New York, 1973), p. 181; *Selected,* I, 33.

12. *Anti-Dühring,* p. 82; Eric Chaisson, *The Origins of Matter and Life: Cosmic Dawn* (New York, 1984), p. 150; *Dialectics of Nature,* p. 194; *Anti-Dühring,* p. 82.

13. On Darwin's materialist outlook see: Stephen Jay Gould, *Ever Since Darwin: Reflections in Natural History* (New York, 1977), pp. 24-25. Engels, *Socialism: Utopian and Scientific, Selected* III, 100.

14. Marx, *ad Feuerbach,* Karl Marx, Frederick Engels, *Collected Works* (New York, 1976) 5:3, 6, 585.*

15. *Selected* III, 101. "In Anfang war die Tat." (In the beginning was the deed!) is from Goethe's *Faust,* part I, Scene 3 (The Study), line 1237. Faust rejects the Biblical "In the beginning was the *Word*" and the Platonic "In the beginning was the Mind" with their theological implications and places the emphasis on natural action.

We might note also the famed occasion on which James Boswell told Samuel Johnson that "it is impossible to refute Berkeley." "I shall never forget the alacrity with which Johnson answered, striking his feet with mighty force against a large stone, till he rebounded from it. 'I refute it *thus.*'" *Boswell's Life of Johnson* (New York, 1908), I, 315 (1763). Johnson instinctively used the criterion of practice, which Kant failed to do when facing Berkeley's subjective idealism.

16. Marx, *Wages, Price and Profit, Selected* II, 54. This work (1865) was left untitled by Marx. When it was first published (1898), Marx's daughter, Eleanor, gave it

*Hereafter as *CW* (arabic numeral).

the title by which it is still best known – *Value, Price and Profit.* Ibid., p. 439.

17. *CW,* 5:8.
18. *Feuerbach, Selected* III, 342; *Anti-Dühring,* pp. 101, 102; *Dialectics of Nature,* pp. 158-159, 339. Collected Works: 84, 520. Albrecht von Haller (1708-1777) was a famed Swiss anatomist and physiologist as well as something of a theologian and poet (*Die Alfen,* one of the earliest 18th century paeans to rugged, mountain beauty.)
19. See Carl Grabo, *A Newton Among Poets* (Chapel Hill, North Carolina, 1930), pp. 167-168; Kenneth Neill Cameron, *Shelley: The Golden Years* (Cambridge, Mass. 1974), p. 391; *Marx, Engels, On Literature and Art,* p. 209; Engels, *Socialism: Utopian and Scientific, Selected* III, 131.
20. *Selected,* II, 97-98.
21. Marx to the Editorial Board of the *Otechestvenniye Zapiski (Notes on the Fatherland),* November, 1877; *Selected Correspondence* (Moscow, 1975), p. 294. Engels (ibid., p. 39) makes the same point in a letter to Conrad Schmidt, Aug. 5, 1890: "All history must be studied afresh, the conditions of existence of the different formations of society must be examined in detail before the attempt is made to deduce from them the political, civil-law, aesthetic, philosophical, religious, etc., views corresponding to them." Karl Marx, *The Poverty of Philosophy* [1847], (New York, 1992), p. 83.

We might note also Marx's comment on Ferdinand Lassalle: "He will learn to his cost that to bring a science by criticism to the point where it can be dialectically presented is an altogether different thing from applying an abstract ready-made system of logic to mere inklings of such a system." (Marx to Engels, Feb. 1, 1858, *Correspondence,* p. 105.)

22. *Selected,* II, 98. The Demiurgos, in Plato's *Timaeus* was the creator of the cosmos and the gods, to whom he delegated power (apparently a derivative variant of the old Sumerian and Indian creation myths).
23. *Selected,* III, 361-362. Marx, *The Poverty of Philosophy,* p. 80.
24. Marx to P.V. Annenkov, Dec. 28, 1846, *Selected,* I, 519; Marx to Johann Baptist Schweitzer, Jan. 24, 1865, *Selected Correspondence,* p. 142; Marx to Engels, Jan. 14, 1858, *Selected Correspondence,* p. 93; Marx, *Afterword to the Second German Edition of the First Volume of Capital, Selected,* II, 98; Engels to Marx, June 16, 1867, *Correspondence,* p. 220: "You ought to have dealt with this part in the manner of Hegel's Encyclopedia, with short Paragraphs, every dialectical transition marked by a special heading . . . " Marx to Engels, June 22, 1867, *Selected Correspondence,* p. 177, Karl Marx, *The Poverty of Philosophy,* pp. 76-85.
25. Marx, *Afterword, Selected,* II, 98.
26. Marx to J.B. Schweitzer, Jan. 24, 1865, *Selected Correspondence,* pp. 144-145, 148. The work Marx wrote "as a reply" to Proudhon was *The Poverty of Philoso-*

phy; see pp. 79, 141.

Friedrich Raumer (1781-1873), a reactionary historian, lawyer and politician who strove for a German Empire under Prussian dominance, was a professor of political science and history at Berlin University when Marx was a student there, at first specializing in law and history. Marx, then, doubtless had not only read some of Raumer's (numerous) works but had heard him lecture. (Raumer, *Encyclopedia Britannica,* 11th ed.)

27. Loren, R. Graham, *Science and Philosophy in the Soviet Union* (New York, 1972), p. 475; G.V. Plekhanov, Appendix, *Fundamental Problems of Marxism* [1892], (Moscow, 1974), p. 89; Engels, *Anti-Dühring,* p. 31; *Socialism: Utopian and Scientific, Selected,* III, 131; *Feuerbach,* ibid., pp. 349-350, 339; V.I. Lenin, *Karl Marx,* in *Marx, Engels, Marxism,* pp. 7-11.

28. *Dialectics of Nature,* p. 26; viii, xii (Preface by Haldane); Gustav Mayer, *Frederick Engels* (New York, 1936), p. 253. On Schorlemmer, *see also* Engels to Marx, June 16, 1867 and May 30, 1873, *Correspondence,* pp. 221, 322.

29. *Dialectics of Nature,* pp. 206-207.

30. *Anti-Dühring,* p. 132.

31. *Dialectics of Nature,* pp. 257, 259.

32. *Anti-Dühring,* p. 133.

33. Marx, Preface to *A Contribution to The Critique of Political Economy, Selected,* I, 502; Engels to Edward Bernstein, Oct. 20, 1888, *Correspondence,* p. 382n.

34. *Dialectics of Nature,* p. 170; Stephen Jay Gould, *The Panda's Thumb, More Reflections in Natural History* (New York, 1980), p. 184.

35. Engels, *Feuerbach, Selected,* III, 349-350, 362, 363. William Butler Yeats, *Among School Children:*

   How O body swayed to music,
   * O brightening glance,*
   How can we know the dancer
   * from the dance?*

36. *Selected,* III, 339; *Dialectics of Nature,* p. 13; Joseph Stalin, "Dialectical and Historical Materialism," *Selected Writings,* p. 407. We might also note the following passage in Engels (*Socialism, Utopian and Scientific, Selected,* III, 129): "Dialectics, on the other hand, comprehends things and their representations, ideas, in their essential connection, concatenation, motion, origin, and ending." (See also *Anti-Dühring,* p. 29.) Again, the initiating process of the interpenetration of opposites is omitted. We might note, too, that in the passage from *Feuerbach* on contradiction quoted above that Engels does not state that contradiction is the essential motion force but only that we "become involved in contradiction" "when we consider things in their motion."

   Stalin's views first appeared in his 1906-1907 essay *Anarchism or Socialism?* (J.V. Stalin, *Works,* I, Moscow, 1952, p. 301.)

37. *Dialectics of Nature,* p. 27.

38. Ibid., pp. 30-31.

39. Ibid., pp. 33-34. We might note that Engels here also implies that nature is essentially developmen-

tal and fails to mention contradiction as the spur to quantity-quality change.

40. Ibid., pp. 29, 27. Lucretius, as I have noted, argued that new phenomena arose from the arrangements of the elements of the old.

41. Notes to *Anti-Dühring* in Appendix 1 to *Dialectics of Nature*, p. 321; see also p. 313. *Anti-Dühring*, pp. 75, 52; see also *Dialectics of Nature*, pp. 263, 30. There was at the time no clear concept of either atoms or molecules. "The atom – formerly represented as the limit of possible division," Engels informed Marx, "is now nothing more than a relation although Monsieur Hofmann himself falls back every other minute into the old idea of actual indivisible atoms." (A.W. Hofmann, *Einleitungin die moderne chemie*, 1866-67.) The confusion is still apparent in the *Britannica* "Molecules" article as late as 1911: "The smallest unit of matter with which physical phenomena are concerned is the *molecule.*" (11th ed., XIII, 655).

42. Plekhanov, *Fundamental Problems of Marxism*, p. 35. We might note that a kind of mystique seems to have developed in regard to rapid changes in biological evolution which are sometimes depicted as inherent in the genetic evolutionary process. They often appear, however, to be triggered rather by chance external (environmental) factors, including weather changes. Biological changes – as in growth – certainly often appear in spurts

but evolution is not simply a matter of biological change.

43. Hegel, quoted in: Raymond Plant, *Hegel* (Bloomington, Indiana, 1973), pp. 142, 137; *Anti-Dühring*, pp. 148-149. In these passages Engels again imaplies that nature is essentially developmental. On the validity of the negation of the negation in regard to society, see, Cameron, *Marxism, A Living Science*, pp. 169, 211.

44. *Dialectics of Nature*, pp. 314, 242, 326. The "law" of the conservation and transformation of energy may, however, have wider application than Engels realized. The process it represents may, for instance, be present in the transformation of photons (of light) and molecules (of smell, taste, touch and sound) by the brain into feelings, behavior and thought (as Lenin seems to suggest, p. 89). The process could also be hidden in the complex patterns of social transformations.

45. *Dialectics of Nature*, pp. 49, 47.

46. *Anti-Dühring*, p. 75; *Dialectics of Nature*, p. 314. Engels' extension of "motion" to history perhaps derived from a comment by Marx in his early (1846-1847) *The Poverty of Philosophy*, (New York, 1992), p. 78: "All that exists, all that lives on land and under water, exists and lives only by some kind of movement. Thus the movement of history produces social relations; industrial movement gives us industrial products, etc." Marx was attacking the (idealist) abstractionist concept of motion in Hegel and

Proudhon and arguing that reality manifests concrete movement. The formulation nevertheless is loose. But Marx goes on to indicate his underlying concept: "the struggle between these two antagonistic elements ["the positive and the negative, the yes and the no"] comprised in the antithesis constitutes the dialectical movement."

47. *Selected,* III: 363, (*Feuerbach*); 128, (Socialism, Utopian and Scientific).

48. *Poverty of Philosophy,* pp. 80-81. *See also,* on categories, *Dialectics of Nature,* pp. 153, 206, 209 and 183: "Abstract identity, like all metaphysical categories, suffices for everyday use, where small-scale conditions or brief periods of time are in question." For more extensive conditions or periods of time, identity, as Engels saw matters, begins to blend with difference.

49. Engels, *Socialism, Utopian and Scientific, Selected,* III, 129: *Dialectics of Nature,* pp. 183, 162, 182.

50. Engels to H. Starkenburg, Jan. 25, 1894, *Correspondence,* p. 518.

51. *Dialectics of Nature,* p. 183.

52. *The Fundamentals of Marxist-Leninist Philosophy* (Moscow, 1974), pp. 162-165; see also p. 131, *Great Soviet Encyclopedia* (New York, 1976), XI, 191. Essentially the same position is presented in other Soviet works, for instance, in A.P. Sheptulin, *Marxist-Leninist Philosophy,* 1978, Yu A. Kharin, *Fundamentals of Dialectics,* 1981, Alexander Spirkin, *Dialectical Materialism,* 1983,

V. G. Afanasyev, *Dialectical Materialism,* 1987. The categories concept is not present in Stalin although it could have some base in his Aristotelian approach to dialectical materialism.

Actually the 50 or so categories in the *Fundamentals* and the Encyclopedia are only a beginning. New categories seem to spring up like weeds. Let us, for instance, consider the essays in *Philosophy in the USSR: The Problems of Historical Materialism,* Moscow, 1981. On page 72, we learn that "mode of life" is a "category," on the next page we have "the category of socio-economic formation." On page 221 "town and country" are described as "categories." On page 239 we have "the category of superstructure." The general approach underlying this plethora of categories is given on page 198:

The task of philosophers in the study of this interdisciplinary problem ["the study of the mechanism of operation of social laws"] is, first and foremost, to develop a methodological basis for these studies, their categorical apparatus, and to provide answers to such questions as "What is the operation of law?"

This is, of course, straight out of Kant: the problem is said to be to uncover the relevant categories, not to find the facts. However, whereas Kant had only twelve categories, the Soviet philosophers had at least sixty.

We might note, too, the linkage of category and law. The

search for categories is rivalled only by that for laws. And here again the field is virtually infinite (p. 171): "The various levels and spheres of being are subject to universal laws expressing the unity of the world, and at the same time, each of these qualitatively specific levels of being has its own specific laws." Thus philosophers have to study "the law of the formation of the well-rounded individual" (p. 199). Historians have to uncover "the laws of superstructure" (p. 239); so that superstructure becomes both a category and a repository of the "laws" of historical development. V. G. Afanasyev in *Dialectical Materialism* (1987) states (p. 96) that "the basic laws of Marxist dialectics ... represent the relationship of connection of categories. ... Hence, without a knowledge of categories it is impossible to comprehend the laws." The "law" of the change from quantity to quality, then, is not a reflection of natural process but a compilation of categories.

That this treatment of categories and laws is idealist, I argued in Appendix II to my *Stalin: Man of Contradiction* (1987, 1989).

There was a similar obsession with "science." Everything is a science. Philosophy is a science, scholarship is a science, dialectics is a science, almost any academic discipline is a science. It is time to restrict the word to its basic meaning. Otherwise any kind of speculation can be presented as scientific.

The approach in all these works, however, is not only idealist but reactionary for it degrades Marxism and separates ordinary people from it. How can workers hope to master so esoteric and unreal a doctrine? Why should they?

The views of the philosophers were not shared by all sections of Soviet society. In 1947 A.A. Zhdanov attacked Soviet philosophers for their elitism and ignorance of science. In 1962 Pyotr L. Kapitsa, a leading scientist, stated that if the Soviet scientists had "paid attention to the philosophers" there would have been no space program. (Richard T. De George, *Patterns of Soviet Thought,* Ann Arbor, Michigan, pp. 188, 208.) In 1987 I was visited by a young Soviet intellectual, who liked my book on Stalin, with its attack on Soviet philosophy. "I often wondered," he said, "where all those categories we had to study in college came from." "Well, one thing is sure," I replied, "they didn't come from Marx, Engels or Lenin." "Yes," he said, "that is certainly true."

53. Marx to P.V. Annenkov, Dec. 28, 1846, *Correspondence,* p. 13; *Poverty of Philosophy,* p. 81.

54. *Poverty of Philosophy,* p. 78.

55. *Anti-Dühring,* Preface, p. 13; Engels, *Feuerbach,* p. 42fn. While there is no reason to doubt Engels' statement that he read "the whole manuscript" of *Anti-Dühring* to Marx, we could wish that he had given some detail. Why, for instance, did he need to read it to

Marx? Why could not Marx read it for himself? Was it all read in one day? Did Marx interrupt the reading and make suggestions? If so, were some of these silently included? Were some suggestions not included or not included properly?

56. H.G. Wells, *The Outline of History* (New York, 1949), p. 1135. Wells' book was written in 1918-1919 and was revised several times thereafter.

57. Engels, *Socialism: Utopian and Scientific, Selected,* III, 103; see also: ibid., pp. 133, 191 (*Origin of the Family*), 335 (Foreword to the 1888 edition of *Feuerbach*); Lenin, *Karl Marx, Marx, Engels, Marxism,* pp. 12-13. See also, Cameron, *Marxism, A Living Science,* pp. 32-34, and Chapter 7, below.

58. In the final paragraph of the Preface to my *Marxism, A Living Science,* I write: "Unlike Marxism, it [dialectical materialism] is not a science but a philosophy and a method of thought." This formulation is wrong because it implies that dialectical materialism is not part of Marxism. Marxism embraces both a social science and a materialist philosophy, separate but interlocking fields.

59. *Fundamentals of Marxist-Leninist Philosophy* (Moscow, 1974), pp. 269, 270; Joseph Stalin, "Dialectical and Historical Materialism," in *Leninism, Selected Writings* (New York, 1942), p. 406, 415.

Chapter Three

1. Engels, *Feuerbach, Selected,* III, 374; Engels, *A Critique of the Draft Social-Democratic Programme of 1891,* ibid., p. 437; *Communist Manifesto, Selected Works,* I, 136. Engels, of course, did not foresee that the bourgeois revolution would continue in a new form as imperialism spread in Asia, Africa and Latin America, blending with anti-feudal, anti-imperialist and pro-socialist forces. Christian churches have often joined these forces, for instance, instigating a "liberation theology" movement. Moreover, in monopoly capitalist countries many religious movements, responding to popular pressures, have to one or another degree taken progressive positions. But on the whole their position has been that of either part of or ally of the State.

2. "The Attitude of the Workers' Party Toward Religion," V.I. Lenin, *Marx, Engels, Marxism* (New York, n.d.), pp. 193, 199, 196. "Fear created the Gods" is from Statius' epic, *Thebais,* on the expedition of the Seven against Thebes.

3. "Religion is the sigh of the oppressed creature, the heart of a heartless world, just as it is the spirit of a spiritless situation. It is the opium of the people." Marx, "Introduction to the Critique of Hegel's Philosophy of Right," (1844) quoted in *Reader in Marxist Philosophy,* ed. Howard Selsam, Harry Martel (New York, 1968), p. 227. The passage was

written before Marx was fully aware of the role of the class struggle in history. He thinks of religion primarily as a needed narcotic and does not note its role in regard to the State.

4. "On the Significance of Militant Materialism," *Marx, Engels, Marxism,* p. 215.

5. "Socialism, Utopian and Scientific," *Selected,* III, 100-103.

6. Loren R. Graham, *Science and Philosophy in the Soviet Union* (New York, 1972), p. 475; G. Plekhanov, *Fundamental Problems of Marxism* (Moscow, 1974), pp. 89, 35.

7. Lenin to A.N. Potresov, June 27, 1899, *The Letters of Lenin,* ed., trans., Elizabeth Hill, Doris Mudie (New York, 1937), p. 90. See also p. 20 fn.

8. Feb. 25, 1908, *Letters,* p. 264; March 24, 1908, V.I. Lenin, *Collected Works, XXXIV* (Moscow, 1972)*, 388. For another translation, see *Letters,* p. 268. Lunacharsky, a writer and playwright, became People's Commissar of Education following the October Revolution.

9. "Materialism and Empirio-Criticism," *CW,* XIV, 51.

10. Ibid., pp. 34, 36, 38.

11. "Socialism, Utopian and Scientific," Marx and Engels, *Selected III,* 101–102; *Feuerbach,* ibid., p. 347; *CW,* XIV, 142–143, 276.

12. *CW,* XIV, 136. Lenin had sometimes to use subterfuge terms such as "fideism" for "religion"

_____
*hereafter, *CW,* (roman numeral).

to get his book past the Czarist censor.

13. Ibid., p. 342.

14. *The Omni Interviews,* ed., Pamela Weintraub (New York, 1984), pp. 21, 28 (Crick), 131 (Pert), 175 (Lilly), 203 (Perry), 325 (Josephson), 362 (Dyson).

15. *CW,* XIV, 252.

16. Ibid., pp. 251, 273, 276, 281, 260; XXXVIII, 57; Lenin, *Marx, Engels, Marxism,* p. 10; "Significance of Militant Materialism," ibid., p. 218. On Soviet philosophers and scientists on Einstein's theories, see Graham, pp. 115, 491 and elsewhere.

17. *CW,* XIV, 260, 281.

18. Ibid., pp. 281, 310; Engels, "Notes to Anti-Dühring" in Appendix 1 to *Dialectics of Nature,* p. 317; Engels to Marx, June 16, 1867, *Correspondence,* p. 221 (see Chapter Two, above, note 41 on the confusion in Engels' day between atoms and molecules).

19. *CW,* XIV, 262. That the theory of an "ether" pervading space was false was first demonstrated by the U.S. physicist Albert A. Michelson in 1881. His experiment was repeated in conjunction with Edward W. Morley in 1887. In 1907 Michelson won the Nobel Prize for physics. But to judge by Lenin's comments the ether theory still continued to have some vogue. Engels, however, had some doubts about it: "If it [the ether] exists at all it must be of a material nature." (*Dialectics of Nature,* p. 258.)

20. Ibid., pp. 346, 312–313.

21. *CW,* XXXVIII, 177, 569; *Letters,* p. 356; *Marx, Engels, Marxism,*

p. 218; *CW,* XXXVIII, 104. Shortly after beginning the Granat article, Lenin was arrested and imprisoned in Galicia (then part of Austria) for about two weeks as a presumed "spy." He then left for Bern where he finished it. In a letter to his sister, A.I. Elizarova, Nov. 1/14, 1914 (*Letters,* p. 347), he writes: "I am finishing my article for Granat's Encyclopedia [about Marx] and am sending it to them one of these days." In a letter to the Secretary of the Granat Publishing House, Moscow, Jan. 4, 1915 (*Letters,* p. 356), he says he would like to make "certain corrections in the section on dialectics" if he could get proofs. He may, then, have made some changes in the article in early 1915 after he began his Hegel studies. (*Marx, Engels, Marxism,* p. 218.) See also *CW, XXI,* 463 and N.K. Krupskaya, *Reminiscences of Lenin* (New York, 1975), pp. 275-283, 295-296.

22. *CW, XXXVIII,* 359, 350 (Philo on Heraclitus), 359. Lenin's apparent quotation in his Note "in the interests of popularization" does not appear to come from Engels. Perhaps it came from Plekhanov or it may simply represent Lenin's summary statement of a current argument. In his *Philosophical Notebooks,* Lenin commented on Ferdinand Lassalle's *The Philosophy of Heraclitus the Obscure of Ephesus,* Berlin, 1858. *CW, XXXVIII,* 341-354.) Marx called Lassalle's book "a very insipid compilation."

(Marx to Engels, Feb. 1, 1858, *Selected Correspondence,* p. 94.)

23. Marx, *The Poverty of Philosophy,* p. 95.

24. C.W. XXXVIII, pp. 359-360. Mao Tse-tung in his 1937 essay *On Contradiction* attempted to develop Lenin's comments on identity and unity, and although he provides new insights, gets tangled up in an eclectic, sometimes non-dialectical, mixture of examples ("Without 'below' there would be no 'above.' ") Mao Tse-tung, *Selected Works,* (New York, 1954), II, 43.

25. *CW,* XXXVIII, p. 177, 360. The conception of development as "self-movement", as I have noted, appears in Stalin's essay *Dialectical and Historical Materialism* and then in later Soviet texts. See Cameron, *Stalin,* pp. 145-146. See also *The Fundamentals of Marxist-Leninist Philosophy* (Moscow, 1974), p. 125, where we learn that "the principle of motion, change and development" is "the universal fundamental principle of all being." So, too, in V.G. Afanasyev, *Dialectical Materialism* (New York, 1987) p. 56: "The development of the material world is an interminable process of the dying off of the old and the emergence of the new." All this, of course, is pure idealism.

26. *CW,* XXXVIII, 222.

27. See *Anti-Dühring,* pp. 28-29; C.W. 25: p. 23; "The Junius Pamphlet," *CW,* XXII, 309. Rosa Luxemburg wrote a pamphlet, *The Crisis in Social-Democracy,* under the pseudonym Junius, with which

Lenin had disagreements. A series of letters signed Junius appeared in *The London Advertiser* in the later 18th century attacking George III and his supporters. It led to a long controversy on the identity of "Junius." Rosa Luxemburg perhaps took the name from this source.

28. Mao Tse-tung, "On Contradiction," *Selected Works,* II, 45. Mao's views have been attacked in a number of Soviet works, for instance, in *A Critique of Mao Tse-tung's Theoretical Conceptions,* (Moscow, 1972). Although this criticism is often sound, it is also sometimes confused. For instance, in the *Critique* (see p. 34), the implication is that opposites do turn into each other, not just that Mao's examples were wrong.

29. *CW,* XXI, 54. See *Marx, Engels, Marxism,* p. 11 for another translation:

> A development that repeats, as it were, the stages already passed, but repeats them in a different way, on a higher plane ("negation of negation"); a development, so to speak, in spirals, not in a straight line; a spasmodic, catastrophic, revolutionary development; "breaks of gradualness"; transformation of quantity into quality; inner impulses for development, imparted by the contradiction, the conflict of different forces and tendencies reacting on a given body or inside a given phenomenon or within a given society; interdependence, and the closest, indissoluble connection between all sides of every phenomenon (history disclosing ever newer and newer sides), a connection that provides the one world-process of motion proceeding according to law.

30. *Marx, Engels, Marxism,* p. 10.
31. Ibid., pp. 11, 52; Cameron, *Stalin,* pp. 146-147, 178-179.
32. *CW, XXXVIII,* 195, 372, 283, 167, 212.
33. V.G. Afanasyev in *Marxist Philosophy* (Moscow, 1980) p. 59, writes "consciousness is inseparable from highly-organized matter and is its product." Yu A. Kharin, *Fundamentals of Dialectics* (Moscow, 1981), p. 81: "Consciousness ... is the product of highly organized matter, the human brain, rather than of all matter." "Consciousness," Kharin continues ... "depends on the appearance of higher forms of motion." This last not only omits the biological factor but is truly confusion worse confounded. Afanasyev continues his basically idealist view in his *Dialectical Materialism* (1985): "all matter contains the intrinsic general property of reflection."
34. *CW, XXXVIII,* 171, 191.
35. Ibid., p. 373: Einstein, quoted in: Heinz R. Pagels, *The Cosmic Code: Quantum Physics as the Language of Nature* (New York, 1983), p. 41; P.B. Shelley, *Hellas: A Lyrical Drama,* (1821) lines 197-200. Einstein also commented as follows: "For the creation of a theory, the mere collection of recorded phenomena

never suffices – there must always be added a free invention of the human mind that attacks the heart of the matter." (Quoted in Pagels, p. 41.)

36. Marx, *Wages, Price and Profit, Selected Works, II,* 54; *CW, XXXVIII,* 153, 130. Lenin was using "moment" in a philosophical sense, used by Hegel to indicate "steps, stages, processes." Ibid., p. 319. See also p. 147.

37. Ibid., p. 193.

38. Ibid., p. 93.

39. Ibid., pp. 94, 256. "Everything," however, as we have noted, does not "develop." It simply changes. Only some changes result in development.

40. Ibid., p. 363. Engels in *Dialectics of Nature* (p. 159) noted that "knowledge develops in a curve that twists many times." Lenin must have created his "spiral" metaphor independently, however, unless this passage or a similar one in Engels appeared somewhere earlier. *Dialectics of Nature* was not published until 1925, a year after Lenin's death.

41. V.I. Lenin, "Once Again on the Trade Unions," *Selected Works* (New York, 1937), IX, 66. Lenin suggested in a parenthesis that all "young members of the Party" study "all that Plekhanov wrote on philosophy." However, although Plekhanov was certainly far in advance of the other socialist intellectuals of his day, his treatment of philosophy (and Marxism in general) is often abstractionist and mechanistic. As one would expect, the basic weaknesses in his political views, especially his waverings on revolutionary action and proletarian power, reflect themselves in his general thinking both in philosophy and social science, for instance, his doctrinaire picture of "the superstructure," which underlay both Bukharin's and Stalin's views on the subject. See my article "The Fallacy of 'the Superstructure,' " *Monthly Review,* XXXI (Jan. 1980), pp. 27-36.

Lenin illustrated his "all sidedness" argument with Bukharin by ingeniously referring to the different uses of a simple drinking glass (perhaps picking one up from the table or lectern in front of him). (*Selected Works,* IX, 65.)

42. *CW,* XXXVIII, 159, 160; Engels, "Socialism, Utopian and Scientific," *Selected,* III, 129. David Hume, as I have noted, was the first major philosopher to question causation, claiming that what we call causation is simply repetition.

43. Note on Dialectical Method.

As I have shown, Lucretius and other materialists perceived dialectical process in nature. Heraclitus saw "strife" as the essence of reality. Without it "all things would pass away." When Lucretius argued that his "multitudinous atoms . . . realized everything that could be formed by their combinations" and that "from their disharmony sprang conflict" he seems on the verge of formulating the "laws" of "contradiction" and qualitative change arising from quantitative and combinational change. When he argued that "sentient" mind emerges from "insentient" atoms he also saw something of the essence of dialectical process. When Diderot argued that perhaps the "vegetable kingdom" arose from the mineral and the animal from the vegetable, he was thinking dialectically, as he was in his discourse on the egg. But although Lucretius, Diderot and Holbach perceived dialectical patterns in natural development, they were alike unable to formulate the generalizations that these patterns indicated, or perceive their fundamental significance.

All three were, of course, hampered in perceiving patterns in natural process by the level of science in their respective ages but this was not in itself an insuperable barrier. Clearly Diderot and Holbach had sufficient knowledge of nature available to have made dialectical generalizations and even in Lucretius' day there was considerable knowledge of nature. The basic problem was not cultural but social. A slave-commercial state was not likely to accept conflict and change as the essence of reality. And even as commercial capitalism advanced to the triumph of the bourgeoisie in the French revolution, and developmental theory arose in all spheres of thought – Kant on the universe or Godwin on society – the only consistently dialectical philosophy that the rising bourgeoisie could give birth to was, as Marx noted, idealist (Hegel). Only with early industrialization do we see, for instance in Shelley, further elements of dialectical materialism emerge. The advance to Marxist social and philosophical views had to await the rounding out of the proletariat with its revolutionary impact on society. This, as we can now see, marked an unprecedented advance not only in society but in human thought. True, the proletariat was bombarded, as the peasants, slaves and artisans had been, by the narcotics of prejudice and superstition, but because it had achieved payment for its work, was independently organized, and operated an integrated economic system it was able to attain literacy and develop the world-view inherent in its exploitive condition much more fully than had those previous exploited classes. It perceived that it could run society without benefit of an exploitive class. "The working class," the Chartist leader, Bronterre O'Brien, noted in 1833 (when Marx was but 15 years of age), " . . . aspire to be at the top instead of at the bottom of society – or rather than there should be no bottom or top at all." In such a class there lay the potential for creating a world-view expanding beyond the limits imposed by the bourgeois need to delude in order to preserve an exploitive society.

When in the hands of Marx and Engels materialism for the first time included the analysis of society as well as of nature, it acquired a new range. And it acquired a new depth also, by combining a view of oppositional interactions as the spur to change with the criterion of practice. When we begin to think dialectically, we see that all things proceed with a back-and-forth wave motion, in which an initial surge forward tends to continue; only when the opposite movement begins to dominate does it fade, at first slowly and then rapidly. We can see this, for instance, as conflicting weather patterns interact, or in recovery from an illness in which we go through an upward curve with ups-and-downs within it, or in a strike or a revolution. So, too, psychologically; for instance, in facing a decision, conflicting emotions and ideas sway back and forth until finally a solution emerges – often swiftly; or in artistic creation, in the shaping of a poem with its amorphous swirls of concepts suddenly taking shape. The same general processes, that is to say, are visible in matter, living matter – as in embryo formation – society, psychological life and cultural creation.

To reflect these and other aspects of reality in our ways of thinking is by no means easy. For instance a present process, social or natural, is often so vivid that we tend to accept it as given, without seeing its interconnections or tendencies; but unless we do, we see it incompletely. It is difficult to realize that everything grew out of something else and that roots, often remote and seemingly improbable, are nevertheless present.

I might note my own experience in materialist-dialectical "method" in writing *The Young Shelley* (1950) and *Shelley: The Golden Years* (1974). I began, as Marx did in *Capital,* with research, putting the material, old and new, together, and then exploring its interconnections. As a Marxist I could not – as the so-called New Critics were doing at the time – artificially narrow the area of relevance to one of examining Shelley's poems simply as aesthetic objects. Nor could I see them, as other bourgeois critics did, as primarily psychological projections; or, as still others did, as pure form – viewed in terms of *Aristotle's Poetics* by the University of Chicago school – or as disembodied "style." Esthetics, form and biographical or psychological matters had, of course, to be examined but they had to be examined in the light of a total picture, one with degrees of foreground and background. Biographical facts had to be related to the social scene. Shelley was not, as Santayana proclaimed, "a finished child of nature" (or of God) but, more mundanely (and complexly) was shaped by his struggle against an exploitive and oppressive order, one that reflected conflicting feudal and capitalist interests. Furthermore, he shared the cultural scene with other writers: Wordsworth, Coleridge, Blake, Byron, Keats, Hazlitt, Hunt, Jane Austen. All reflected an age in turmoil. What were the social roots of their ideas and emotions? What views did they have in common? In what respects did they differ, specifically how did Shelley differ? Certain common social forces – essentially conflictive class forces – had molded all of them whether they knew it or not. Moreover, these forces were ultimately rooted in the new productive forces and productive relations of the industrial revolution – Blake's "dark Satanic Mills" – that had churned up the society from its foundations. Unless things are seen in their dependencies, they seem simply chaotic. It was because these writers were shaped by common social forces that they shared similar general patterns of thought and style. Sometimes these forces were reflected directly, as in Blake's

The Chimney-sweeper or Shelley's *Song to the Men of England,* but more often indirectly as in Shelley's *Ode to the West Wind* with its metaphorical vision of historical change, or in his *To a Skylark* with its soaring exaltations of freedom. They were reflected also in the fluid intensity of the style of both poems –

> *Drive my dead thoughts over the universe*
> *Like withered leaves to quicken a new birth!*

– a style directly opposite to Alexander Pope's feudal-formalistic 18th century heroic couplet with its pony-trot cadence and social subservience.

> *In spite of Pride, in erring Reason's spite,*
> *One truth is clear, WHATEVER IS, IS RIGHT.*

It might be argued that all these factors do not need to be taken into account; but if, like Lenin, we seek allsidedness or depth of understanding they do, particularly so in considering poets like Dante, Lucretius, Byron, Shelley, Goethe, or Neruda, who reflect and enrich the broader social streams of thought in their age. Unless we perceive that Shelley's *Prometheus Unbound* is basically a revolutionary drama, depicting the feudal-capitalist struggle and the future rise of an egalitarian world, we are, to paraphrase a 19th century critic on Shelley, like someone writing an essay on Vesuvius without noting that it is a volcano. Yet it is not only a revolutionary drama. It also reflects aspects of Shelley himself in his identification with the rebel Prometheus and it contains philosophical and scientific as well as social significance.

Chapter Four

1. On the atom see: Gerald Feinberg, *What is the World Made of? Atoms, Leptons, Quarks and other Tantalizing Particles,* New York, 1978; Nigel Calder, *The Key to the Universe,* Penguin Books, 1978; Heinz R. Pagels, *The Cosmic Code: Quantum Physics as the Language of Nature* (New York, 1983), pp. 178-181, 221; John Gribbin, *In Search of the Big Bang,* New York, 1986.

2. Feinberg, p. 73.

3. Marcia Bartusiak, *Thursday's Universe* (New York, 1988), pp. 225-226; *New York Times,* Oct. 13, 1989 (on leptons consisting of quarks).

4. John Gribbin, *The Omega Point* (New York, 1988), p. 82.

5. Bartusiak, pp. 225-226; Calder, p. 50.

6. Gribbin, *Big Bang,* pp. 7-11 (Wright and Kant); Bartusiak, pp. 174-176, 120. The light waves recorded at the red end of the electromagnetic spectrum are long, those at the blue end are shorter. John Gribbin and Martin Rees, *Cosmic Coincidences, Dark Matter, Mankind and Anthropic Cosmology* (New York, 1989) p. 236. *New York Times,* Jan. 16, 1991 (quasar clusters).

7. Bartusiak, 93-104; Gribbin, *Omega Point,* pp. 31, 116-1117, *New York Times,* June 10, 1993.

8. Bartusiak, 107-109.

9. *New York Times,* Feb. 28, 1989, Jan. 16, 1991, June 3, 1991.

10. Bartusiak, p. 129.

11. Steven Weinberg, *The First Three Minutes* (New York, 1977), pp. 44-51. Gribbin, *Big Bang,* p. 138; Bartusiak, 119-120. Gribbin and Rees, pp. 63-99. Late in 1989 a quasar was discovered 15 billion light years away, the furthest away object yet detected. Hence it is 15 billion years old, also an argument for the 20 billion year age estimate of the universe. (*New York Times,* Nov. 20, 1989.) See also *New York Times,* Nov. 22, 1992, indicating that "a generation of stars" had been born and died by 12 billion years ago.

12. Bartusiak, pp. 242-245; Gribbin, *Omega Point,* pp. 49-50.

13. Gribbin, *Omega Point,* p. 35; Bartusiak, pp. 215; *Omega Point,* pp. 82-84; *New York Times,* Jan. 24, 1989.

14. Bartusiak, p. 180, *New York Times,* Feb. 16, 1988 (quoting Tully); Bartusiak, pp. 199-201; Gribbin, *Omega Point,* pp. 150-153; *New York Times,* Jan. 24, 1989; Gribbin and Rees, pp. 63-99. *New York Times,* June 8, 1993 (Milky Way "halo"). The "dark matter" theory, although in recent years attracting more and more adherents, is viewed skeptically by some. (*New York Times,* Jan. 16, 1991.)

15. *New York Times,* Feb. 28, 1989, Dec. 24, 1993 (satellite evidence); Bartusiak, p. 216; Gribbin, *Omega Point,* pp. 38-39, 83-84; *New York Times,* Jan. 24, 1989, Jan. 14, 1990 (a report that a spacecraft late in 1989 failed to find irregularities in the very early universe). *New York Times,* April 17, Dec. 2, 1992.

16. Gribbin, *Big Bang,* pp. 175-177; Bartusiak, pp. 281-282 (quoting Gerald Share); Gribbin, *Big Bang,* p. 178; *New York Times,* April 27, 1993 (oxygen and carbon).

17. Bartusiak, pp. 244; 261-262, 264; Gribbin, *Big Bang,* pp. 372-377.

18. Bartusiak, p. 263; Gribbin, *Big Bang,* pp. 375-376. The "singularity," "vacuum fluctuation" and similar speculations about the beginning before the beginning are most prominently associated with Edward P. Tryon and Stephen Hawking, the author of the metaphysical *A Brief History of Time.* For a summary of Hawking's views, see Gribbin, *The Big Bang,* pp. 378-392. For Tryon and others, see Bartusiak, pp. 253-273.

19. For Lucretius, see Chapter One.

20. Bartusiak, p. 225 (quoting Weinberg).

21. Francis Crick, *What Mad Pursuit, A Personal View of Scientific Discovery* (New York, 1988), p. 40.

22. J. D. Bernal, *Science in History* (Cambridge, Massachusetts, 1965), III, 986, 987; Mahlon B. Hoagland, *The Roots of Life* (New York, 1977), pp. 35-39; *New York Times,* Oct. 13, 1989.

23. *New York Times,* Aug. 12, 1980; Bernal, p. 896.

24. Calder, p. 19.

25. Hoagland, p. 15.

26. S. E. Luria, *Life, The Unfinished*

Experiment (New York, 1973), p. 100. (In the 1940's one of my closest friends at Indiana University was Salvadore E. Luria, as he notes in his autobiography (*A Slot Machine, A Broken Test Tube,* New York, 1984). Hence I had a kind of layman's inside track as the science of molecular biology was developing, and felt the excitement of unfolding discovery. It was then, as Luria reported enthusiastically on his "phage" experiments, that I first learned that not protein but something unknown before, namely the nucleic acids, underlay heredity. One of Luria's students at the time was James Watson, whom I met then, and later at Cold Spring Harbor on Long Island after his and Crick's double helix discovery.

27. Engels, we might note, states that qualitative change can come about either by "addition" or "subtraction" of "matter or motion." Frederick Engels *Dialectics of Nature* (London, 1940), p. 27. *New York Times,* Feb. 12, 1991 (Michaelson); March 20, 1990 (retinoic acid).

28. Only certain parts of the x and y chromosomes, it has been found, contain the sex-determinant genetic material. *New York Times,* May 29, 1983; Dec. 23, 1987.

29. Quoted in Stephen Jay Gould, *The Panda's Thumb: More Reflections in Natural History* (New York, 1982), p. 64.

30. *New York Times,* Nov. 9, 1982; May 8, 1990.

31. Eric Chaisson, *Cosmic Dawn: The Origins of Life and Matter* (New York, 1984), pp. 171-173; John Gribbin & Jeremy Cherfas, *The Monkey Puzzle* (New York, 1983), pp. 20, 22. Bees are thought to have split off from wasps some 125 million years ago. (*New York Times,* Dec. 8, 1987.)

32. George Gaylord Simpson, *The Meaning of Evolution* (Mentor Books, 1951), p. 25.

33. TV film, *The Flight of the Condor,* part 2, U.S. Public Television, 1980 (Andes); Simpson, pp. 66, 76; *New York Times,* Dec. 8, 1987.

34. Simpson, p. 34; *New York Times,* Dec. 8, 1987.

35. Chaisson, p. 178. Alfred Tennyson, *Locksley Hall,* 1,137.

36. John Gribbin, *In Search of the Double Helix* (New York, 1987), pp. 210-211. Luria, pp. 19-20.

37. Gribbin, pp. 68-69; Crick, p. 25.

38. Sewall Wright, quoted in G. Ledyard Stebbins, *Darwin to DNA, Molecules to Humanity* (San Francisco, 1982), p. 146; Gribbin & Cherfas, pp. 53, 121. On these various matters see also P. B. Medawar and J. S. Medawar, *Aristotle to Zeus, A Philosophy Dictionary of Biology,* Cambridge, Mass., 1983.

39. Donald Johanson & Maitland Edey, *Lucy: The Beginnings of Humankind* (New York, 1981), p. 250; Richard E. Leakey, *The Making of Mankind* (New York, 1981), pp. 40-42; Johanson & Edey, pp. 274-275; John Gribbin & Jeremy Cherfas, *The Monkey Puzzle* (New York, 1982), p. 182; Richard E. Leakey and Roger Lewin, *People of the Lake: Mankind and its Beginnings* (New

York, 1978), pp. 119-120: Gribbin & Cherfas, pp. 200-201, 239, 29; John Gribbin, *Double Helix,* pp. 343-346. *New York Times,* Nov. 18, 1993, on the discovery of the Lucy subpeople males ("found three years ago.") Bernard Campbell, *Human Evolution* (Hawthorne, NY, 1975) p. 30, on great apes' body weight.

40. Johanson & Edey, pp. 99, 102; Carl Sagan, *The Dragons of Eden: Speculations on the Evolution of Human Intelligence* (New York, 1977), p. 94; Leakey, *Making of Mankind,* pp. 113-125, 118; Johanson & Edey, p. 103: Grahame Clark, *World Prehistory, A New Outline* (Cambridge, England, 1969), p. 13; *New York Times,* Feb. 18, 1988; Dec. 1, 1988. On these datings and other matters, see also *Ancestors: Four Million Years of Humanity,* American Museum of Natural History, New York, April 13-September 9, 1984, and Richard Leakey and Roger Lewin, *Origins Reconsidered,* New York, 1992. Whether the Neanderthals died out or blended into still existing human stocks is still being debated. (*New York Times,* Feb. 4, 1992.)

41. *New York Times,* Dec. 14, 1993, on human evolution and climate change. On the Australian natives, see the Epilogue in Gribbin and Cherfas (pp. 247-258) which presents some interesting research along with some apparently minimally supported speculation.

42. Engels, "The Part Played by Labor in the Transition from Ape to Man," Marx and Engels, *Selected,* III: 66; Gould, p. 132; Gribbin, pp. 343-346.

Chapter Five

1. *Anti-Dühring,* p. 42; Engels, *Feuerbach* (Appendices), p. 67.

2. Marx and Engels, *Selected,* I, 13; Lenin, "Materialism and Empirio-Criticism," *CW,* XIV, 51; Lenin, *Philosophical Notebooks, CW,* XXXVIII, 167.

3. *The Brain and Psychology,* ed. M.C. Wittrock (New York, 1980), pp. 9-10; William H. Calvin and George A. Ojemann, *Inside the Brain: Mapping the Cortex, Exploring the Neuron* (Mentor Books, New York), p. 13.

4. Wittrock, pp. 14-15.

5. Jay Braun, Darwyn E. Lindner, *Psychology Today, An Introduction* (New York, 1979), pp. 57, 58; Carl Sagan, *The Dragons of Eden, Speculations on the Evolution of Human Intelligence* (New York, 1977), p. 32 (citing experiments by Wilder Penfield); Wittrock, pp. 9, 14.

6. Sagan, pp. 41-42; *The Omni Interviews,* ed. Pamela Weintraub (New York, 1984), p. 121 (Candace Pert); Eric Chaisson, *Cosmic Dawn, The Origins of Life and Matter* (New York, 1984), p. 151; *The Brain and Psychology,* pp. 3-4; Sagan, p. 47.

7. William H. Calvin and George A. Ojemann, p. 4; Sagan, p. 34; Richard M. Restak, *The Brain, The Last Frontier* (New York, 1979), pp. 297, 93; *New York*

Times, Aug. 18, 1987; Sagan, pp. 31-32.

8. *New York Times,* Dec. 6, 1989 (on pigeons); Penfield, quoted in Restak, p. 230; See also, on specialized brain function, Calvin and Ojemann, pp. 25, 29-30, 153-154. (In 1931, when, as a student I was preparing to leave McGill University for Oxford I sat next to Wilder Penfield at a Rhodes Scholars' dinner. He talked about Pavlov, in whom I was then particularly interested, and whose work had influenced him, and then, to my surprise, about Edna St. Vincent Millay, particularly her sonnet sequence, *Fatal Interview,* which he knew in detail. He told me that he kept her poems beside his bed and read them every night.)

9. George Johnson, "Memory: How It Works," *New York Times, Magazine,* Aug. 9, 1987.

10. Restak, p. 123; Calvin and Ojemann, pp. 10, 18, 112; Restak, pp. 124, 117; Calvin and Ojemann, p. 122; *New York Times,* July 6, 1993.

11. Isaac Asimov, *The Human Brain, Its Capacities and Functions* (Mentor Books, 1963), pp. 271-272, 302; *Harper Collins Dictionary of Biology,* (New York, 1991).

12. Asimov, pp. 139-140, 271-272, 141-142; *Comparative Psychology,* ed. M. Ray Denny (New York, 1980), p. 417; Hugh Brown, *The Brain and Behavior* (New York, 1976), p. 18.

13. Denny, p. 299. There are also mating rituals, of a simpler nature, among insects. See for instance, Stebbins, p. 95.

14. N.J. Berrill, *Sex and the Nature of Things* (New York, 1955), p. 119.

15. Sagan, p. 38; Denny, p. 197.

16. Denny, pp. 199, 200; Asimov, *The Brain,* p. 321; Engels, "The Part Played by Labor in the Transition from Ape to Man," *Selected,* III, 73-74; Edward O. Wilson, *Sociobiology* (Cambridge, Mass., 1980), p. 88.

17. On the banana experiment, done by Wolfgang Kohler and recorded in his *Mentality of Apes* (1925), see *Psychology Today,* p. 127. Gribbin & Cherfas, p. 221.

18. Sagan, pp. 116-117. For some criticism of the ape language experiments, see Gribbin & Cherfas, pp. 213-219. *New York Times,* June 25, 1985 (on Kanzi); Leakey, *People of the Lake,* p. 156.

19. O. Yakhot, *Materialist View on Reality* (Moscow, n.d.), pp. 88-89. Yakhot calls the experimental animal a monkey. Other descriptions of the experiment call it an ape, which seems more likely. See *ABC of Dialectical and Historical Materialism* (Moscow, 1976), p. 118.

20. Wilson, p. 289.

Chapter Six

1. Dec. 12, 1868, Karl Marx and Friedrich Engels, *Correspondence,* 1846-1895 (London, 1939) p. 255.

2. Jay Braun, Darwyn E. Linder, *Psychology Today, An Introduction* (New York, 1979), pp. 380-383; William H. Calvin, George A. Ojemann, *Inside the Brain*

(New York, 1980), p. 194; *New York Times,* March 25, 1980; Edward O. Wilson, *Sociobiology* (Cambridge, Mass., 1980), pp. 78, 123-126; Richard M. Restak, *The Brain, The Last Frontier* (New York, 1979), p. 228; Jo Durden-Smith, Diane deSimone, *Sex and the Brain* (New York, 1983), pp. 99, 123-124. *New York Times,* July 16, 1993 (on the genetic base for male homosexuality. Research on female homosexuality is underway.) The female hormone, progesterone, is produced in the corpus luteum. Estrogen is also produced, in small quantities, in the testes.

3. Engels to Conrad Schmitt, Nov. 1, 1891, *Correspondence,* p. 495.

4. Actually only a small part of the genetic material in the y chromosome determines sex. Presumably the same is true for the x chromosome but this has not yet been determined. *New York Times,* Dec. 23, 1987.

5. *New York Times, Science Times,* May 8, 1990. (On experimental evidence supporting "sexual selection"); Durden-Smith and deSimone, pp. 81-84, 131-132.

6. Restak, pp. 223-227; Durden-Smith and deSimone, pp. 67-68, 39-141, 146-147, 152, 246-247.

7. Durden-Smith and deSimone, pp. 87-88; *The Omni Interviews,* ed. Pamela Weintraub (New York, 1984), p. 129.

8. Durden-Smith and deSimone, pp. 111-112; *New York Times,* April 11, 1989.

9. *New York Times,* April 11, 1989, quoting Dr. Sandra Witelson who conducted the research; Pert, *Omni,* p. 128.

10. *New York Times,* Nov. 23, 1981; Arlene Elsen Bergman, *Women of Viet Nam* (San Francisco, 1975), pp. 155-157; Kenneth Neill Cameron, *Stalin, Man of Contradiction,* (Toronto, 1987), pp. 80-81. 11, Durden-Smith and deSimone, pp. 213-214; *New York Times,* April 11, 1989.

12. For twenty years, 1843-1863, Engels and Mary Burns lived together as husband and wife in Manchester. After her death, he lived with her sister, Lizzie (Lydia), whom he married on her deathbed. Engels described Lizzie as coming from "real Irish proletarian stock," with "an ardent inborn feeling for her class." She was also, like her sister, an Irish nationalist revolutionary. She died in 1878. Engels did not write *The Origin of the Family, Private Property and the State* until 1884, but its feminism contains insights and passion that could have had its roots in his relationship – over 25 years – with these two radical Irish women. Little appears to be known about Mary and Lizzie Burns. It seems a fruitful field for research. For some general facts see *Frederick Engels: A Biography* (Moscow, 1974), pp. 212, 217, 235, 307, 337. An aversion to a marriage ceremony was not uncommon among both men and women radicals at the time. They especially resented state interference in personal relationships, a view which perhaps had some roots in Shelley's *Queen Mab* (1813), part of which

Engels translated into German. For some account of Engels' views on women and society see my *Marxism, A Living Science,* (New York, 1994). The continuing influence of working class and Marxist thought on the British feminist movement was vividly depicted in the television series *Shoulder to Shoulder,* shown on Public Television in the United States in the 1990s.

Chapter Seven

1. George B. Schaller, *The Year of the Gorilla* (Chicago, 1964), pp. 111, 198-199.
2. Ibid., p. 187.
3. Ibid., p. 187.
4. Ibid., pp. 135-136.
5. Ibid., pp. 35-36.
6. Dian Fossey, *Gorillas in the Mist* (Boston, 1983), pp. 50, 61, 68-70, 150, 190, 98. Dian Fossey was murdered in December 1985, perhaps by poachers, from whom she had long protected the gorillas. The motion picture, *Gorillas in the Mist* (Warner Brothers, 1988; MCA, Home Video, 1989), gives an interesting fictionalized version of Fossey's experiences but omits the main significance of her book, namely her detailed study of gorilla society as a guide for reconstructing the roots of human in hominid society.
7. Ibid., pp. 12, 81-82, 74, 86, 72, 174, 96, 175.
8. Jane van Lawick-Goodall, *In the Shadow of Man* (Boston, 1971), pp. 28-29.
9. Ibid., p. 113.

10. Ibid., pp. 200, 71, 125, 152-153.
11. Ibid., pp. 121, 73.
12. Ibid., pp. 52-53. Many of the scenes described in *In the Shadow of Man* were also recorded on film by Jane Goodall and her husband, a professional film maker.
13. See Samuel K. Lathrop, *The Indians of Tierra del Fuego,* New York, 1928; Richard E. Leakey and Roger Lewin, *People of the Lake: Mankind and its Beginnings* (New York, 1979), pp. 90-114 (on the Kung people). See also Ronald M. Berndt and Catherine H. Berndt, *The First Australians,* New York, 1954. For the parallels between the Pygmy chimps and the Australopithecines, see John Gribbins, *In Search of the Double Helix* (New York, 1987), pp. 344-346. Charles Darwin, *The Expression of the Emotions in Man and the Animals,* London, 1872. On facial expressions in the apes, see Goodall, pp. 273-275. The ways of the Kung People and other hunter-gatherers have fortunately been recorded on film.
14. Berndt and Berndt, pp. 49-50.
15. *A Very Strange Bird . . . The Cuckoo, Geo, The Earth Diary,* III (May, 1981), p. 118; on turtles: Edward O. Wilson, *Sociobiology, The Abridged Edition* (Cambridge, Massachusetts, 1980), p. 29; on wolves: ibid., pp. 246-247; *New York Times,* Jan. 10, 1984.
16. For the gene as God see, for instance, David Barash, *The Whisperings Within,* Penguin Books, 1979. Barash's genes seem to be endowed with consciousness; for

instance, on p. 133, it is theorized that when a parent saves a child from harm, "he is not really an altruist, since his genes are doing neither more nor less than saving some of themselves."

17. Karl Marx, *The Economic and Philosophical Manuscripts of 1844,* ed. Dirk J. Struik (New York, 1973), p. 181; Marx, *Capital, I* (Moscow, 1965), p. 609. The *Capital* passage was quoted by Ernesto Rodriguez in a review of Richard Levin, Richard Lewontin, *The Dialectical Biologist* in *Science and Nature* (Spring, 1989, Nos. 9/10), p. 183. On the early hominid's direct descent from the apes, see John Gribbin and Jerman Cherfas, *The Monkey Puzzle* (New York, 1982), p. 29, and John Gribbin, *Double Helix,* pp. 343-346.

18. Frederick Engels, *Ludwig Feuerbach, Appendices* (New York, 1941), p. 67; Frederick Engels, "The Part Played by Labor in the Transition from Ape to Man" *Selected,* III, 74.

19. *New York Times, Science Times,* May 8, 1990 (experimental evidence on mating instincts).

20. *The Omni Interviews,* ed. Pamela Weintraub (New York, 1984), pp. 128, 129. Candace Pert is a brain research scientist at the Biological Psychiatry Branch of the National Institute of Public Health in Bethesda, Maryland. Engels to Conrad Schmitt, Nov. 1, 1891, Karl Marx and Frederick Engels, *Correspondence,* 1846-1895 (London, 1939), p. 495.

21. Marx in a letter to Johan Baptist Schweitzer on Proudhon wrote of "the simple moral sense which always kept a Rousseau, for instance, from even the semblance of a compromise with the powers that be" (in contrast to Proudhon with his egoism and "vanity"). *Selected Correspondence* (Moscow, 1975), p. 148. Marx, then, believed that there was a "moral sense" that inclined people to act for the good of humanity. He would doubtless – in view of his beliefs in instincts and evolution – have accepted the notion that this sense had evolutionary roots and could be molded in various ways by social forces.

22. For a modern social-determinist view with quasi-Marxist trimmings, see: R.C. Lewontin, Steven Rose and Leon J. Kanin, *Not in Our Genes: Biology, Ideology and Human Nature,* New York, 1984. For other such views, some presented by proclaimed Marxists, see the articles in *The Sociobiology Debate,* ed., Arthur L. Caplan, New York, 1978. These writers dismiss, without examination, the extensive findings of animal society research as a set of "moral tales" (*Not in Our Genes,* p. 160) and simply omit the field in their own presentations.

23. Lenin cites Karl Marx, "Author's Preface, A Contribution to a Critique of Political Economy," in *A Handbook of Marxism* (New York, 1935), pp. 543-44 (a translation I favor). For the ways in which these views were perverted by Stalin and others into a rigid "the base," "the superstructure" theory, see my *Marxism, A Liv-*

ing Science (New York, 1994),
pp. 20-28, 199, and my *Stalin,
Man of Contradiction* (Toronto,
1987; England, 1989), pp. 112-
116.

Chapter 8

1. J.B.S. Haldane, *The Marxist Philosophy and the Sciences* (London, 1938), p. 106; William Blake, "The Garden of Love," in *Songs of Experience.*
2. *The Works of Plato,* ed. Irwin Edman (Modern Library, New York, 1928), pp. 135, 141, 137.
3. Lucretius, *The Nature of the Universe,* trans., R.E. Latham (Penguin Classics, 1951), pp. 124, 126; William Ellery Leonard, *Two Lives, III,* xxi, x.
4. Engels to Friedrich A. Sorge, March 15, 1883, Marx and Engels, *Correspondence, 1846-1895* (London, 1934), pp. 414-415; *Karl Marx, A Biography* (Moscow, 1973), p. 254; Marx to Engels, Sept. 7, 1864, *Correspondence,* p. 158; *Frederick Engels, A Biography* (Moscow, 1974), p. 212.
5. Engels, *Feuerbach, Selected,* III, 345. I have been unable to find the exact quotation in Shaw but remember reading it some years ago. I wrote to Dan Lawrence, my erstwhile colleague at New York University and editor of the letters of Shaw, and he was unable to find it either, although he found similar passages. Engels, we might note, wished his body to be cremated. His ashes were scattered at sea, by Eleanor Marx and three others. (Yvonne Kapp,

Eleanor Marx, II (New York, 1976, 597-599.) Marx realized that the "opium" of religion had deep roots: "Religion is the sigh of the oppressed creature, the sentiment of a heartless world, as it is the soul of soulless conditions. It is the opium of the people." The comment came in Marx's early essay, "A Criticism of the Hegelian Philosophy of Law." See Cameron, *Marxism, A Living Science,* pp. 164, 210. John Keats, *The Fall of Hyperion,* 1. 157.

6. In this connection we might note the following from Mao Tse-tung:

> Death awaits all men but its significance varies with various persons. The ancient Chinese writer Szuma Ch'ien said: "Although death befalls all men alike, in significance it may be weightier than Mount Tai or lighter than a swan's down." In significance, to die for the interests of the people is weightier than Mount Tai, but to work hard and die for the fascists, for those who exploit and oppress the people, is lighter than a swan's down. Comrade Chiang Szee-teh died for the interests of the people, and his death is indeed weightier in significance than Mount Tai.

7. Lewis Thomas, *The Lives of a Cell* (New York, 1979), pp. 114-115; *New York Times,* Dec. 19, 1985 (on animal mass suicides); *Boswell's Life of Johnson,* ed. Chauncey Brewster Tinker (New York, 1933), I, 394.
8. A. E. Housman, *On Wenlock Edge.*

9. "Such is My Will that Thou Thyself go to Eternal Death. In Self Annihilation. . . . ", William Blake, *The Ghost of Abel*. Blake also used the phrase in *Milton, The Four Zoas* and other works.

10. William Wordsworth, *Expostulation and Reply* (1798).

11. *Mary Shelley's Journal,* ed. Frederick L. Jones (Norman, Oklahoma, 1947), July 28, 1814 (entry by P.B. Shelley).

12. George B. Schaller, *The Year of the Gorilla* (Chicago, 1964), pp. 254-258.

13. V.I. Lenin, *Selected Works,* III (New York, 1936), 353; X (1938), 127, 144.

14. Joseph Stalin, *Leninism, Selected Writings* (New York, 1942), pp. 464-465.

15. Ali of Basra led a slave and peasant revolt in Iraq and Iran in the 9th century. See Cameron *Humanity and Society,* p. 240. Hung Hsiu-chuan led the Taiping Rebellion in 19th century China directed against feudal lords and foreign imperialists. The rebels established a new state with Nanking as its capital.

16. P.B. Shelley, *Ode to the West Wind* (1819): Karl Marx, *Critique of the Gotha Program,* 1875 (New York, 1933), pp. 44-45. Shelley used the symbols of winter and spring specifically in a social sense in other works, one of which, *The Revolt of Islam,* contains what is, in effect, an earlier version of the Ode – as his revolutionary heroine, Cynthna, speaks to her companion, Laon, after the defeat of the revolution:

    This is the winter of the world;
    — and here
    We die, even as the winds of
    Autumn fade,
    Expiring in the frore and foggy
    air.
    Behold! Spring comes, though
    we must pass who made
    The promise of its birth . . .

    See Kenneth Neill Cameron, *Shelley: The Golden Years* (Cambridge, Mass., 1974), pp. 293-295.

17. O'Brien, quoted in Cameron, *Marxism, A Living Science,* p. 15; Karl Marx, *Critique of the Gotha Programme,* 1875, (New York, 1933), pp. 44-45.

18. Candace Pert, *The Omni Interviews,* ed., Pamela Weintraub (New York, 1984), p. 131. The comment comes as a bit of a shock after reading Pert's insightful scientific analysis of the human brain.

19. P.B. Shelley, "To a Skylark;" Algernon Charles Swinburne, "Itylus."

20. Dylan Thomas, "The Force That Through the Green Fuse Drives." As Thomas was influenced by dialectical materialism it can be assumed that he was aware of the philosophical implications of his image here. I was part of a panel discussion with Thomas at Indiana University in the late 1940's and was impressed and surprised by his clearly Marxist-influenced radicalism.

21. Edna St. Vincent Millay, "Elegy"; Shelley, "Adonais," lines 179-180 (1821). On a broader death theme, Shelley's wife, Mary Shelley, the author of *Frankenstein* published an end-of-the-world

novel, *The Last Man,* in 1826, which envisioned the human race dying from a plague. See also Byron's vivid 1816 poem, "Darkness," on the perishing of the planet due to natural causes:

The bright sun was extinguished, and the stars
Did wander darkling in the eternal space.

22. Lucretius, *The Nature of the Universe,* trans., R.E. Latham (Penguin Classics, 1951), pp. 183–184.

23. Kenneth Neill Cameron, *Marx and Engels Today: A Modern Dialogue on Philosophy and History,* (Hicksville, NY, 1976), p. 14; Marx, "ad Feuerbach," *CW* 5:p.5.

INDEX

Afanasyev, V. G., 220, 221, 224, 225
agnosticism, 50, 52, 87
Ali of Basra, 203, 228 n.15
Alomaeon of Croton, 23
Anaxagorus, 17
Anti-Duhring, 45-6, 214-5, 221
Antigone, 165
Aquinas, Thomas, 3, 26-7, 29, 72
Archimedes, 18, 19
Aristarchus of Samoa, 23
Aristotle, 17-18, 19, 23, 24, 28, 30,
    32, 37, 48, 72, 76
asceticism, 16, 24
Athens, 15, 16, 19, 190
atoms, 11-12, 29, 47, 118; &
    Democritus, 18; & Lucretius,
    20-22; Engels on, 96, 208, 219
    n.41; Lenin and, 97
australopithecus, 141, 142, 177

Bacon, Francis, 27, 28-9, 34
Bacon, Roger, 3, 27
Bantusiak, Marcia, 120, 122
Barnum, P.T., film on, 197-98
Barrett, Elizabeth, 166
Becquerel, Henri, 94-5
Berkeley, Bishop George, quoted,
    31; 34, 37, 51; & subjective
    idealism, 89; on death, 194; 216
    n.15
Bernal, J.D., 98, 117
Berndt, Ronald M. and Catherine H.,
    178
"Big Bang" theory, 124-25, 126
Blake, William, 3, 193, 207
Born, Max, 123
Boswell, James, 197
Bukharin, Nikolai, 113; 226 n.41
Burns, Lizzie, 165, 234 n.12

Burns, Mary, 165, 196, 234 n.12
Byron, 203, 239

Calder, Nigel, 121, 123, 229
Capital, 27, 55, 56, 57, 58, 217
    n.24
Carlile, Richard, 42, 214 n.1
categories, 36; & Kant, 37; & Aristotle,
    76; discussed 73-79; Marx & Lenin
    on, 112; in Soviet philosophical
    work, 77, 112, 220-21 n.52
causality, 32-33, 114
cell, biological, 132-4, 135; nerve,
    145-46; Engels on, 47, 80, 129
Chartists, 42, 44
China, 13, 14, 23, 24, 105
Cicero, Marcus Tullius, 20
city-states, 15, 25, 190
Civil War in France, The, 80
Clytemnestra, 165
Coleridge, Samuel Taylor, quoted
    191-92; on death, 193-94, 197
commercial capitalism, 30, 40, 43
    105
Communist League, 44
Confucianism, 3, 14
Copernicus, 12, 27
Crick, Francis, 93, 130, 139
Cromwell, Oliver, 33, 196, 203
Curie, Marie and Pierre, 94, 191

Dalton, John, 36
Darwin, Charles, 35, 50, 136, 153,
    207;
    On the Origin of Species..., 45
    48, 136, 181; on *Capital,*
    215 n.8
Darwin, Erasmus, 35
Davy, Sir Humphrey, 36

241